Encyclopaedia

of

NANOSCIENCE TECHNOLOGY AND ENGINEERING

Encyclopaedia of Nanoscience Technology and Engineering

Prof. Pallab Bhattacharya

ANMOL PUBLICATIONS PVT. LTD.
NEW DELHI-110 002 (INDIA)

ANMOL PUBLICATIONS PVT. LTD.

Regd. Office: 4360/4, Ansari Road, Daryaganj,
New Delhi-110 002 (India)
Ph.: 23278000, 23261597

Branch Office: No. 1015, Ist Main Road, BSK IIIrd Stage
IIIrd Phase, IIIrd Block,
Bangalore-560 085 (India)
Tel.: 080-41723429
Visit us at: www.anmolpublications.com

Encyclopaedia of Nanoscience Technology and Engineering

First Edition, 2009

ISBN 978-81-261-4115-9

PRINTED IN INDIA

Printed at Balaji Offset, Delhi.

Contents

Preface

The field of nanoscience has witnessed a rapid growth in the last decade. Recently, the attention of the community of nanoscientists has been focusing more and more on technological applications. Nanotechnology is an enabling technology, with high potential impact on virtually all fields of human activity, yielding high expectations for solutions to the main needs of society, although having to address open issues with respect to sustainability and compatibility. The fields of application of research in nanoscience include: aerospace, defence, national security, electronics, biology and medicine.

There has been significant progress in the understanding achieved in recent years, both from the theoretical and experimental point of view, along with a strong desire to assess the current state-of-the-art in this fast-growing field, stimulating, at the same time, research collaboration among the different stakeholders in the area of nanoscience and the corresponding technological applications, and possibly prompting the organization and presentation of joint projects in the near future, involving both industry and public research. Nanoscience and nanotechnologies involve studying and working with matter on an ultra-small scale.

The technology stretches across the whole spectrum of science, touching medicine, physics, engineering and chemistry. Scientists are looking at whether nanotechnology could be used to improve the delivery of cancer fighting drugs and are examining whether nanoscale carbon could be used to increase the power and speed of computer circuits.

Author

Chapter 1

Nanotechnology

The word nanotechnology is a combination of the two terms "nano" and "technology. "From this alone, it is apparent that nanotechnology is an application of scientific concepts at the level of molecules and particles, the sizes of which fall within the range of 1 to 100 nanometers. In a nutshell, nanotechnology refers to functional engineering at the level of molecules. Specifically, it is the deliberate application of production processes and technology at the molecular level in the order to achieve a desired result, to manufacture materials and consumables with unique characteristics.

Applications of nanotechnology are wide-ranging and far-reaching since it enables humans to make useful products and devise efficient manufacturing processes in any field. As things stand, nanotechnology has made inroads in consumer products like computer mice, mats, storage bags and containers, air purifiers, wound dressings, golf clubs, sunscreen, protective and stain-resistant clothing, computer chips, and others.

Nanotechnology, the word, was first used in 1974 by a Japanese professor to refer to the process of material fabrication at the nanoscale or at the molecular level. The more popular definition of the word, however, was made by American engineer K. Eric Drexler (who was not aware of the earlier definition of the word) to refer to his conception of an "assembler machine" with the capability of replicating itself, and building ever larger and more complex items.

Over the years, however, the definition of nanotechnology has expanded. Its present application refers to any work done at the molecular level so it now covers a variety of technologies and scientific disciplines that are used to manufacture devices and materials at the nanoscale.

Molecular manipulation gained a following due to the distinct characteristics of matter at the nanoscale. It became apparent from scientific studies that matter can exhibit peculiarly different characteristics at the nanoscale. For instance, surface area-to-volume ratios are higher in the microscale than it is in the macroscale.

At the nanoscale, physical characteristics that are so weak or insignificant at the macro-level become more distinct and much more pronounced. To illustrate, copper which is opaque at the macroscale is actually transparent in the nanoscale. Another example is gold which is apparently liquid at the nanoscale whereas it is solid at the macroscale. There are a lot more such reversals when one moves from the macroscopic to the microscopic perspective.

This is one commonly used process of molecular manipulation, where molecules are arranged in a preset or pre-defined manner to come up with materials that would have unique characteristics. One application of this process can be seen at the manufacture of Metal Rubber - a material that has the conductive characteristics of metal but the elastic properties of rubber. In molecular self assembly, manufacturers or engineers supply the molecules of different materials and let them interact and bond naturally (as they would in nature) to form supramolecular materials.

Assembly can be intermolecular (between and among different molecules) or intra-molecular (between and among molecular components).

There is a mystique associated with nanotechnology, but it is actually an outgrowth of existing activities, such as microelectronics and ultrafine particle technology, that involve efforts to engineer materials to achieve specific results. Nanotechnology is a continuation of the trend toward smaller, faster, better. As the scale of materials is reduced, their properties may be enhanced. Materials smaller than 100 nm may have unique properties.

One of the developments over the past few years has been the ability to predict and control these properties. According to the International Organization for Standardization (ISO), nanotechnology involves:

- Understanding and control of matter and processes at the nanoscale, typically below 100 nm in one or more dimensions, where the onset of size-dependent phenomena usually enables novel applications.
- Utilizing the properties of nanoscale materials that differ from the properties of individual atoms, molecules, and bulk matter, to create improved materials, devices, and systems that exploit these new properties.

NANOSCALE

Nanotechnology covers a broad range of fields, from biotechnology to computer chips. Many interesting kinds of nanotechnology involve unusual physical phenomena that only show up at the nanoscale—things like quantum electron effects that can make gold change colour. Nanotechnologists have been searching for these effects, and sometimes turning them into products, for quite a few years now. But they are difficult to find, to study, to understand, and to apply.

Another family of nanotechnology involves nanoparticles used as structural materials. Carbon nanotubes have been used to reinforce tennis rackets; flakes of clay only a few atoms thick have been used to make plastic more leak-proof and fire-resistant. The nanomaterials are typically inert, and often provide only incremental improvement over larger and less engineered materials.

A third kind of nanotech involves putting molecules together in ever more creative ways. Chemists have been building molecules for centuries, but new molecules require intricate recipes developed through trial and error. Each step of the recipe wastes some of the inputs to create unwanted chemicals, and so the recipes cannot include very many steps. To compensate, the chemists have developed an amazingly broad library of techniques to choose from.

There are many other kinds of nanotechnology: for example, modern computer circuits are small enough to qualify, and there's some very exciting work going on in medicine. But in this essay, we will focus on molecular manufacturing, a kind of nanotech that falls somewhere between nanomaterials and molecules.

Molecular manufacturing proposes to build nanoscale mechanical systems, for the purpose of mechanically guiding the chemical reactions that will build more nanoscale mechanical systems.

A few reactions have already been carried out under direct mechanical control, using relatively huge machines—scanning probe microscopes and micro-imprinting stamps. If the machines can be made a lot smaller and more capable, then they may be able to build molecules into structures—even physical duplicates of the tools used to build them.

Tools being used to build duplicates of themselves is not a new concept; a blacksmith could copy just about every tool in his shop using only the tools already there. But there are some new things in molecular manufacturing: precision, automation, scale, and performance. Together, these factors could make molecular manufacturing as revolutionary as the computer or the steam engine.

Atoms such as carbon like to bond to each other in specific arrangements. Pull an atom slightly out of place, and it will spring back. A particular bonded configuration of atoms is called a molecule, and two molecules with the same structure will be effectively identical. A component that is manufactured as a single molecule will have exactly the same dimensions as its copies, unless an atom is grossly out of place forming a noticeable flaw. The difference between manufacturing tolerances and molecular precision is like the difference between a gallon of milk and a carton of eggs: there's never precisely one gallon in the jug, but it's easy to tell if an egg is missing.

When it comes to manufacturing, this precision is highly important, because it means that (unlike all manufacturing techniques to date) a copy of a copy of a copy of a component can be just as precise as the original, without any need for measuring. Even if the fabrication operations are not 100per cent precise (which they will not be, thanks to thermal noise), if they are close enough, then the newly placed atoms will spring into their proper positions.

Precision should make automation easier. Not only are all parts built alike, but they will not suffer wear or fatigue the same way metal and plastic parts do. A molecular machine that is distorted by moderate stress will simply spring back unharmed, without ever suffering fatigue or ablation. A machine could do a huge number of operation cycles without changing its behaviour in the slightest. Using precise mechanical components also implies an ability to exclude contaminant molecules absolutely.

Human technology is just beginning to build machines at the nanoscale. A few motors have been built, out of metal-encrusted nanotubes, out of painstakingly-designed molecules, or out of components copied from biology. A very few mechanisms have been built—things like levers and wheels. But nanoscale machines are a very attractive target for several reasons. Once we can build machines intricate enough to carry out computer-controlled manufacturing operations, we will be able to build a much broader range of machines.

In many ways, small precise things work better than larger things. Performance of nanoscale machines benefits from a few basic facts of physics called "scaling laws. "One of the scaling laws is operation frequency: small machines take less time to do an operation than larger machines. In fact, the operation frequency is inversely proportional to the size: if we could shrink a machine a millionfold, to a scale of nanometers instead of millimeters, it would do a million times as many operations per unit time. Functional density scales with volume: a millionfold shrinkage would lead to a quintillionfold increase in components per volume.

Calculations show that power density should scale with operation frequency: a millionfold smaller machine should have a million times the power density of its larger cousin. It would have a quintillionth of the volume, and thus only a trillionth of the power. But connect a trillion of the tiny machines in tandem, and we'd have the same total power in one-millionth the volume. Of course, such designs would have to be reliable beyond today's dreams—but remember that molecular machines will not suffer wear. And if one of them did happen to break, say from random background radiation, performance would decrease by one-trillionth.

The final calculation from scaling laws is the one that makes all this possible. Atoms are extremely small. A modern scanning probe microscope, trying to build itself one atom at a time, might take six quintillion years. But if that microscope could be shrunk, then not only would the number of atoms required shrink very rapidly, but the microscope itself could work more rapidly.

The combined effect is that the relative throughput—the time required for a machine to make its own mass of product—shrinks as the fourth power of the size. That means that a 100-nanometer scanning probe microscope could, in theory, duplicate itself in 100 seconds.

In addition to scaling law benefits, there are two other sources of performance improvement. One is that flawless molecular structures can be far stronger than macroscale materials. Carbon nanotubes have a theoretical strength 100 times that of steel. The other source of performance comes with the tongue-twisting name of superlubricity, meaning that under certain conditions, flawless flat surfaces can slip over each other with extremely low friction—and no lubrication required.

From atoms, to molecules, to machines, to manufacturing systems that can duplicate themselves using simple molecular feedstocks: this overview has explained only the broadest outline of the approach and its merits, and has not touched on the problems of design, control, and reliability. Those problems appear to be solvable with ordinary research and development. Other problems are somewhat more esoteric and less well understood: for example, it is not yet known exactly how to mechanically manipulate small molecules to make them join into larger functional structures.

There are some practical problems that arise at the nanoscale (including many quantum effects and thermal noise) that will complicate the design process. No one has said that developing this technology will be easy.

There are, of course, skeptics who warn that the practical problems may be insurmountable, or that there may be unexpected theoretical limitations. Further research will tell. The quality of skepticism has improved over the past few years, as molecular manufacturing captures more interest; as a result, a number of rather flimsy skeptical arguments have been abandoned. Fifteen years after the major theoretical work in the field was published, no one has yet identified a factor that would preclude the building of high-performance kilogram-scale products composed of nanometer-scale machinery, using a few hours of time in a kilogram-scale factory. The implications of this capability would be transformative, disruptive—perhaps terrifying—and will be the subject of several future columns.

As skepticism recedes, what is left is a wide-open opportunity. Molecular manufacturing will not replace other forms of nanotechnology—there are too many useful phenomena waiting to be uncovered at the nanoscale. But for straightforward mechanical work, molecular manufacturing may be the most effective approach, with the highest power density and efficiency. Mechanical operations at the nanoscale can implement sensors and computers as well as motors. And the mechanical approach, relegating nanoscale complexities to the lowest levels of structure and preserving precision at all higher levels, should contribute substantially to simplicity of design.

In January of this year, a group of about two dozen scientists got together in Britain for the purpose of inventing research projects that could lead to software-controlled molecular fabrication. Although the scientists involved in this "Ideas Factory" did not in general share the vision presented here, they nevertheless applied themselves to the goal of developing research projects that could implement computer-controlled fabrication at the molecular scale.

They developed three proposals to be funded and completed within just a few years: one to implement a sort of artificial DNA-programmed ribosome, one to manipulate and join molecular structures under precise mechanical control, and one to understand the detailed workings of scanning-probe chemistry.

In this author's opinion, successful completion of these projects will take us a large fraction of the way toward a general-purpose molecular manufacturing capability.

PREDICTIONS FOR MARKET GROWTH

In the United States, the National Nanotechnology Initiative (NNI) reports that federal funding for nanotechnology R&D has increased from $464 million in 2001 to $1. 05 billion in 2006. Globally, government spending on nanotechnology was estimated at $4. 8 billion in 2005. ISO expects the global nanotechnology market to reach $29 billion by 2008. It is estimated that by 2015, nanotechnology will have a $1 trillion impact on the global economy and will employ 2 million workers.

NANOTECHNOLOGY EFFORTS

Nanofabrication may be conducted either in a top-down or bottom-up approach. In the top-down or miniaturization approach, large materials are reduced to a functional nanoscale structure whose performance is critically dependent on having dimensions less than 100 nm. The top-down approach is represented by the semiconductor or microelectromechanical systems (MEMS) sensor world, where photolithography is used to etch patterns onto silicon surfaces to make circuits. Top-down operations are typically performed in tightly controlled environments because they require very precise tools and the product during processing may be sensitive to contamination.

The bottom-up or self-assembly approach starts with molecules and particles and assembles them into larger structures to make useful products. For example, DNA processes are being applied to replicate natural phenomena in which clusters of atoms self-assemble into more complex entities. The biology community does much of this work in liquid phase and it is uncertain to what extent cleanrooms will be used. However, bottom-up technology uses electron microscopy, which requires very stable conditions with respect to vibration.

Role of IEST

IEST has defined a series of nanotechnology-related working groups, which currently operate as a single unit to chart the path of the organization's nanotechnology initiative. These working groups build on IEST's background in developing Recommended Practice (RP) standards for cleanrooms and controlled environments.

Essentially, the areas of interest are the specification, design, construction, and operation of facilities for nanotechnology R&D and manufacturing; terminology and protocols related to environmental parameters for these facilities; and testing of products manufactured using nanotechnology.

Developing an initial guidance document for facilities is a high priority. We hear about the large, high-profile nanofacilities, such as the new National Institute of Standards and Technology (NIST) centre or the major university laboratories. Many of these expensive facilities incorporate extensive cleanrooms and vibration control techniques. Many organizations, however, are unable or unwilling to invest that level of resources. Therefore, IEST has the opportunity to address the question, what does it take to do nanotech work? Can less-expensive facilities be designed? How can existing facilities be adapted? Requirements will vary.

For research purposes, much can be accomplished with small cleanrooms and gloveboxes. Manufacturing requires more extensive facilities. Startup companies with limited venture capital may need to start with modest facilities they will soon need to expand as the organization grows. IEST guidelines would benefit organizations seeking an economical way to become involved in nanotechnology. IEST has some members who are very experienced in building the new generation of nanofacilities and who can contribute to the development of these guidelines.

Many nanotechnology processes may require new approaches to controlled environments. In addition to developing new guidelines, it will be important to look at updating existing RPs to include nanotechnology considerations.

This will affect documents issued by the Contamination Control Division as well as the Design, Test, and Evaluation/ Product Reliability Division. One possible issue is the control of molecular contamination affecting the product or process. With smaller and smaller particles, more stringent cleanliness levels may be needed in certain applications.

ENVIRONMENTAL AND SAFETY CONCERNS

Personnel exposure to nanoparticles and other nanomaterials during manufacture is an important issue. In the document *Approaches to Safe Nanotechnology,* the National Institute for Occupational Safety and Health (NIOSH) notes that in the absence of scientific data, the degree of risk is unknown because the characteristics of nanomaterials may be different from those of the larger particles with the same chemical composition. As new information becomes available, NIOSH will continue updating guidelines for monitoring, engineering controls, personal protective equipment, and exposure limits.

In addition, various standards organizations are addressing environmental and toxicity questions in their nanotechnology committees. IEST working groups will consider the industrial hygiene policies and practices that are being developed and cite the appropriate resources in the Recommended Practices. However, health, safety and environmental considerations are still at a very early stage. As nanotechnology evolves there may be industry-specific needs that IEST could fill.

IEST has a unique niche to fill. There is a demand for guidelines, recommended practices, and standards for developing nanofacilities. This is a tremendous opportunity to extend what we know about controlled environments and environmental testing to the nanotechnology field. At the same time, it is essential to coordinate with other organizations, keep abreast of their activities, and avoid duplication of work.

For example, the Institute of Electrical and Electronics Engineers (IEEE) recently published a standard focused on carbon nanotubes. ASTM International has a strong focus on toxicology and environmental impacts.

International Electrotechnical Commission (IEC) efforts are centered on electricity and related technologies. ISO is concentrating on nomenclature, measurements, and environmental health and safety. IEST currently sits on the US Technical Advisory Group for ISO Technical Committee (TC) 229 Nanotechnologies and can provide input and review of the US positions. Further, ISO/TC 209 Cleanrooms and controlled environments (IEST holds the Secretariat) is in the process of establishing mutual liaisons.

Nanotechnology is still in its early phases. IEST and other organizations must follow the progress of research results into the manufacturing arena. As developments take shape, IEST can build on its current base of standards and recommended practices. The important task at hand for IEST is to develop the first round of guidance documents specifically for nanotechnology application. IEST members can make important contributions to this megatrend that promises to affect everything in our lives.

NANOETHICS

Nanotechnology is the study, design, creation, synthesis, manipulation, and application of functional materials, devices, and systems through control of matter at the nanometre scale (1-100 nanometres, one nanometre being equal to 1×10^{-9} of a meter), that is, at the atomic and molecular levels, and the exploitation of novel phenomena and properties of matter at that scale.

The beginnings of the field can be dated relatively precisely to the 1959 speech by Richard Feynman entitled

"Plenty of Room at the Bottom", but it is only in the last 10-15 years that some kinds of deliberately produced nanotechnology have reached practical use (e. g. Pilkington's self-cleaning window panes).

Alongside the practical development of nanotechnology ethical considerations have also begun to emerge, and it has been suggested that we need a new kind of ethics "nanoethics" to handle the ethical problems inherent in nanotechnology. Is this a reasonable suggestion? It is difficult to specify exactly what could make an area of technology so special that it needs its own ethics, but a minimal requirement must be that it either raises ethical issues that are not raised by other kinds of technologies, or that it raises ethical issues of a different (i. e. larger) magnitude than other technologies. Is this the case for nanotechnology?

THE NANO-DYSTOPIANS

On one side of the debate about the implications of nanotechnology we find a group of commentators who hold a view which can be called "nano-dystopian". The development of nanotechnology will create great risks for society, and may in the end lead to the destruction of the human race. In the UK this view has gained prominence because it has been actively promoted by prince Charles.

The most "popular" doomsday scenario among nano-dystopians is the "grey goo" scenario in which selfreplicating nano-robots get out of control and devour all matter turning the earth into a seething mass of grey goo. A slightly more refined version of this scenario in which the nano-robots only devour organic material forms the basis for Michael Crichton's 2002 novel "Prey".

Certain other uses of nanotechnology have also attracted particular negative attention. These are the use of nanotechnology for surveillance and the possible convergence of nanotechnology with a range of other technologies in a "project" eventually leading to human enhancement.

Nanosurveillance can be achieved by passive tagging of people or products with nanotransponders that can be read by an interested agent at some distance with the appropriate reader (in reality the transponders are currently and in the foreseeable future considerably larger than nanoscale); or it could perhaps in the future be achieved by active surveillance nanomachines. Other worries have focused not on the technology itself but on the forces that are shaping its development trajectory. It has for instance been noted that a lot of nanotechnology research is performed for and sponsored by the military.

THE NANO-UTOPIANS

On the other side of the debate we can find nano-utopians who predict that nanotechnologies are going to solve all our problems, and make us live for ever. It is on this side of the debate we find predictions about medical nano-robots that will clean up our clogged arteries from the inside, and generally repair all our ills, while other nano-robots will work tirelessly to clear up the environmental mess of the last 300 years of industrial society.

This nano-utopian view has been promoted by the American writer Erich Drexler and his foresight institute since the 1980s. A specific strand of nano-utopianism focuses on a possible future convergence between nanotechnology, information technology, gene technology and cognitive science that will. This convergence is predicted to be the breakthrough that will finally allow effective cognitive enhancement of human beings (and presumably animals as well), and it is therefore highly popular among trans- and posthumanist thinkers.

A PLEA FOR NANO-REALISM

Neither the dys- or the utopian predictions are plausible, and even if they were they would be at time scales much longer than those often bandied about. Predictors of technological developments seem to have an unfortunate attraction to either a 3-5 or a 5-10 year timeframe for the successful development of a new technology, and they are often completely resistant to revision of these estimates. But it is completely unlikely that we will see, for instance, functioning medical nano-robots within the next 10 years. This does, however, not mean that nanotechnology is not extremely important.

The use of nanomaterials in consumer products does, for instance, raise new toxicological and environmental challenges. Some nanomaterials are completely new and have never before existed in nature. This is for instance the case for nanotubes and fullerenes.

Their toxicological and environmental characteristics are therefore also unknown. But there are also new issues concerning nanosized particles of known materials. Because of their size nanoparticles are dispersed differently in the environment and can gain access to other sites in the body than larger particles can, and partly because of their very large surface area to mass they can have different chemical properties than larger scale particles of exactly the same material.

This means that even though we know a certain chemical to be "safe", this may no longer hold if it is formulated in nanosize particles. Health and safety issues related to nanomaterials and nanoparticles are however, still in an early stage and, as a consequence, it seems premature to draw far-ranging conclusions regarding the potential hazards and risks related to exposure to these materials. The available data regarding the health hazards and risks of exposure to nanomaterials and nanoparticles are limited and there are no systematic investigations.

Recent data demonstrate that lung exposure to nanoparticles (<100 nm) produce greater adverse inflammatory responses compared with that of larger particles of the same composition and at equivalent mass concentrations 7. The interaction of particles with cells and the consequence of surface area properties and free radical generation play an important role in nanoparticle toxicity.

This effect is due to the high sizespecific deposition of nanoparticles when inhaled as single particles rather than aggregated particles. It was concluded that surface area might play an important role in nanoparticle toxicity. It should be noted that size effects play an important role not only for the intensity of potential adverse effects (in the case of toxic materials), but also for the efficiency of nanoparticle-based drug delivery systems.

NEED OF SPECIFIC NANOETHICS

The strongest case for the need for a specific nanoethics could be made if either the predictions of the nano-dystopians or the nano-utopians are correct. So, let us for the sake of the argument assume that these predictions are correct, and that nanotechnology has great destructive and other worrying potentials or great beneficial potential (or perhaps even both at the same time). Is that enough to require us to create nanoethics as a separate field of inquiry? Probably not.

There are plenty of extant technologies that could destroy the earth, or at least destroy the earth as we know it (e. g. atomic technology, the burning of fossil fuels), plenty of technologies that may be used for intensifying surveillance of citizens, and plenty of technologies where development is partially driven by military needs.

At the same time the positive claims made for nanotechnology has over time been made for a range of other technologies, most recently in the medical field gene technology and stem cell therapies. While there is thus a great need for ethical analysis of the many ethical issues raised by the different developments in nanotechnology, there is no need for a specific nanoethics. The toolbox developed in applied ethics during the last 35 years probably already contains the necessary tools for the analysis of nanotechnology.

DEVELOPMENT OF MATERIALS

Nanotechnology involves development of materials (and even complete systems) at the atomic, molecular, or macromolecular levels in the dimension range of approximately 1-500 nanometers. Current research looks to provide detailed understanding of unique properties that materials exhibit at the NanoScale.

Current focus of the research is positioned to create and

use structures, devices, and systems that have unique and often contradictory properties as well as enhanced functions because of their small and/or intermediate size.

Nanotechnology research and development includes control at the NanoScale and integration of NanoScale structures into larger material components, systems, and architectures as well as automated systems for the production of NanoMaterials and the automatic assembly of structures and systems.

Nanotechnology first gained recognition after Nobel Laureate, Richard Feynman, presented his talk, "There's Plenty of Room at the Bottom" to the American Physical Society in 1959. Activity surrounding Nanotechnology began to slowly increase over the next few decades. In 1988, Eric Drexler taught the first course on Nanotechnology.

In that programme, he suggested the possibility of nanosized objects that were self-replicating. Slow growth in this area of scientific investigation continued. The next major milestone was when Rice University Professor Richard Smalley won the 1996 Nobel Prize for discovering a new form of carbon: a molecule of sixty carbon atoms (referred to as C60). Today C60 has become one of a growing number of building blocks for a new class of nanosized materials.

The advancements in Nanotechnology really began to accelerate in the late 1990s. Nanotechnology captured the thoughts and imagination of scientists and venture capitalists after an entire issue of Scientific American addressed the promise of this new technology.

CURRENT STATE OF NANOTECHNOLOGY

Research into NanoMaterials spans a significant spectrum of areas. Advanced material companies are producing innovative products in areas such as coatings, industrial powders, chemicals, and carbon nanotubes. Today, real world application of Nanotechnology exists in commercial business.

About two-dozen serious applications of NanoMaterials and process have been fielded ranging from non-scuff floor tile to high strength brackets for running boards on vehicles to high temperature protective materials for spacecraft. While NanoMaterials are a significant portion of today's focus, several other areas are equally as promising.

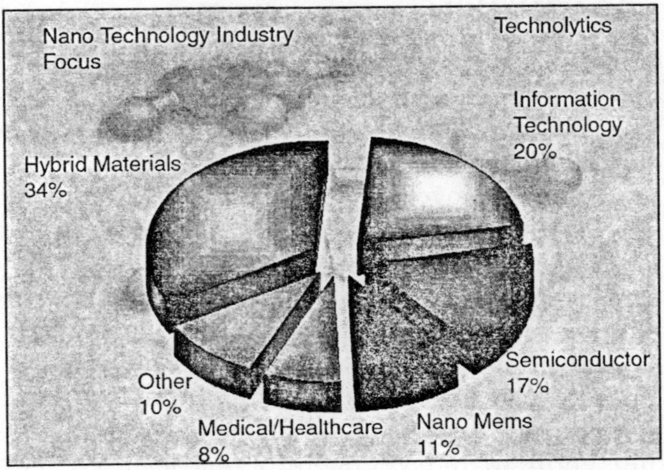

Fig. Current State of Nanotechnology

TRENDS & IMPACT OF NANOTECHNOLOGY

Nanotechnology has the potential to become a more significant revolutionary force for business than the industrial revolution or the information technology revolution. In fact, many believe that the combined impact of both the industrial and information revolution may approach the magnitude of change that could result from the commercialization of Nanotechnology.

Currently, Nanotechnology is moving from the basic research stage of its evolution into the applied research stage of technology maturity.

Today there are several Nanotechnology companies already being traded on the public marketplace. As this technology evolves and matures, we can expect to see many more companies enter this space. Today's manufacturing methods are very crude at the molecular level. Casting, grinding, milling, and even lithography move atoms in mass. It is like trying to make things out of LEGO blocks with boxing gloves on our hands. Yes, we can push the LEGO blocks into great heaps and pile them up; but we cannot really snap them together the way we would like.

FUTURE ADVANCES

- Self-assembling products, systems and materials (self repairing system)
- Computers billions of times faster (reaching the size barrier of microprocessors)
- Extremely novel inventions (self-adapting fabrics)
- Realistic space travel and exploration (more economic)
- NanoMedical (unique treatment delivery capabilities)
- Molecular food syntheses (end of famine and starvation)

The big question is "When will this technology enter into the mainstream?" As the graphic below indicates, it will take another eight to ten years. But, rest assured, several breakthroughs and advances will produce significant advantages in materials science, semiconductors, optical communications, and information technology.

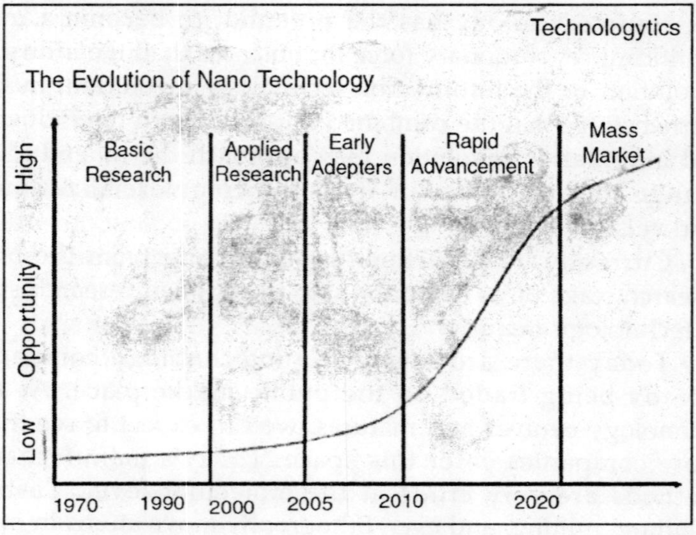

Fig. The Evolution of Nanotechnology

HOMELAND SECURITY APPLICATIONS

With the broad reach Nanotechnology has in terms of capabilities, the direct applications for Defence and Homeland Security are only limited by our imagination and how rapidly the technology advances. From shape-shifting armor to fabric that can turn away microbes as well as bullets to new power sources, the defence industries are launching major initiatives and planning for Nanotechnology.

The Government is the major source of funding for current Nanotechnology initiatives. Centers of Excellence in Nanotechnology have been established around the country. The basic research in Nanotechnology conducted at these centers will provide the foundation upon which real world applications can be built. Other centers are already concentrating on military application of Nanotechnology. While there are efforts for new and improved weapons based on Nanotechnology, the vast majority of the Nanotechnology research and applied research fall into the support category.

From improved powers sources and batteries to advanced arming fuses the defence industry and homeland security has great interest in how Nanotechnology their capabilities. They believe that Nanotechnology will advance sensors and protective equipment and will greatly assist them in their mission. In fact, current research and development efforts are working on micro power generating devices that can be used in a plethora of applications.

If successful these MPG can provide enough energy to power sensors and sensor networks that will be deployed to protect critical infrastructure like water treatment plans, roadways, and bridges. Other applications could include such things as microscopic self-powered reconnaissance and surveillance devices like listening devices, vibrations sensors as well as supplying power to sensor networks. Currently, DoD is funding research in small-scale energetic device development and management.

The lifecycle development and support requirements for NanoMEMS (micro-electro-mechanical systems) in support of advanced applications in munitions and armament systems are critical to advances in weaponry. NanoMEMS has direct implications on integrated circuits, optical switches, pressure sensors, as well as the processes used to produce weapons. However, current thoughts are that NanoMEMS can play a major role in new weapons and intelligence and surveillance. Imagine using NanoMEMS to construct a steam powered electrical generator systems for satellites. Steam is currently used in conventional and nuclear power plants to turn turbines that generate electricity.

When used onboard a satellite, as the satellite spins and out of direct sunlight, the steam condenses to liquid; and when it rotates back in the sunlight, the liquid turns to steam again. The micro-size steam power plant would generate enough electricity to operate the unit. The satellite's overall demand for energy would have been greatly reduced because of the use of extremely efficient electronics created with Nanotechnology.

In addition, the overall size of the units could be reduced by a factor of ten or more, as well as the weight because of construction practices at the NanoScale. Just think! A satellite so small that it could be launched by what appears to be a large gun in place of conventional rockets. The cost savings would be enormous as well as the difficulty in anyone being able to detect the satellite in orbit.

The area that Nanotechnology that holds the most significant promise is in the area of sensors and sensor networks. Hybrid NanoMaterials will produce orders of magnitude improvements in high-selectivity and high-sensitivity sensors for biological and chemical detection. This advanced detection of harmful chemical and biological agents; microsensors for radioactivity, low-power or self-powering consumption security electronics, polymer electronics, nano-optics will provide capabilities that are not available today.

Just imagine. In a few years advanced sensor networks self powered with the smarts to communicate from sensor to sensor and have the ability to detect very small amounts of chemicals or biological agents installed in the water supplies across the country. Once a single sensor node detects the presence of one of the agents, it communicates to the others what hit was found and receives their verification.

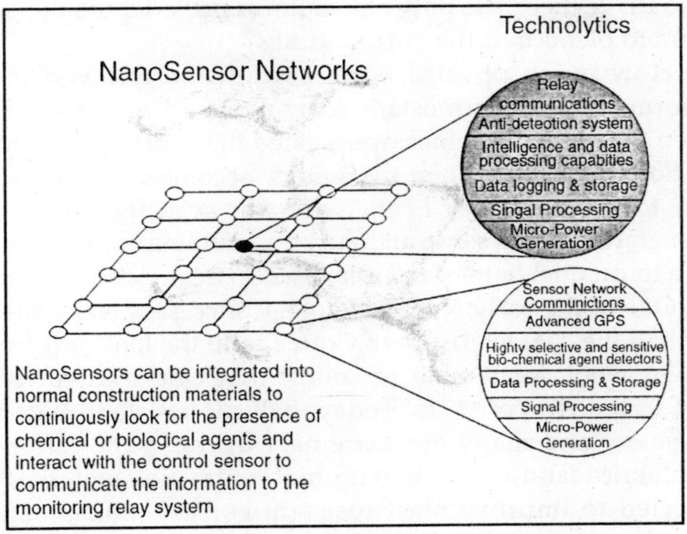

Fig. Nanosensor Networks

Once verified, the information is communicated to the control sensor that relays the information back to the National Infrastructure Protection Centre for immediate action. Other applications like underwater sensor networks to detect the movement of ships into and out of our ports could also be advanced using Nanotechnology. Sensor networks that detect chemical, biological or radiological materials could be built into cargo containers.

The applications of NanoMaterials go far beyond semiconductors and sensors into NanoFabrics. NanoFabrics with unique properties are under development at this time. Properties like decreased receptivity to chemical or biological agents, materials with the ability to expand and contract (like a thermostat) as to exhaust or conserve body heat, or to resist the penetration of a bullet.

Consider that today a soldier going into battle carries about 60 lbs of equipment. A significant part of that weight is in the bullet -proof-vest. What if the weight burden was reduced by 50 per cent? How much more efficient would the soldier be? What if instead of having multiple types of camouflage there was only one uniform that adapted the outer colors to blend into the surroundings?

How much physical stress would be relieved if the uniform had that thermostatic characteristic discussed earlier? When the soldier becomes over heated the vents open to allow the flow of air and when the soldier becomes cold the vents close to conserve body heat. There was even the notion that the materials could sense an injury and automatically constrict like a tourniquet or mast trousers used to treat victims in shock.

One of the early Nanotechnology successes was material that had the characteristics of Gortex with the look and feel of regular wool. Remember of course that Gortex is currently used in bulletproof vests. Today, ballistic resistant materials are heavy and many are extremely brittle. Super strength nanofabrics sandwiched into normal contraction materials are expected to improve blast resistant construction practices making commercial and governmental structures much more resilient against bomb blasts.

Bomb resistant containers for cargo and luggage on ships and aircraft, bomb resistant glass for office buildings and government complexes, advanced structural members that have the strength by are pliable to absorb the energy of a blast are all currently being investigated as real world application of Nanotechnology.

Another area of Nanotechnology that is receiving the greatest amount of attention is semiconductors. The ability to further compact the number of transistors in a given space increases the performance of a semiconductor. NanoScale construction practices applied to semiconductors will substantially increase current processing capability and could change the entire industry overnight.

Nanotechnology also has significant benefits in opto-electronics and communications. The ability to construct an optical switch on a chip would eliminate a significant amount of complexity and cost of optical networks, not to mention increasing the capacity as well.

While Nanotechnology is based in the research labs today, the advances made to date have illustrated the significant value this technology will bring. The implication on manufacturing techniques, materials manufacturing and the semiconductor industry will be profound. The enhanced characteristics of materials will allow us to create new and innovative devices to protect all of us from terrorism, both directly through safer construction designs and indirectly through intelligence and surveillance.

With each passing day, the promise of Nanotechnology becomes increasingly apparent. There is a long way to go before the NanoMaterials production techniques can provide adequate supply of high quality materials at affordable prices. This is one technology that requires close monitoring to properly time the significant opportunities that will be created in it wake.

When Nanotechnology will be in its mature form it is sure it will have its impact upon almost every industries land almost every area of society from communication to medicine, from agriculture to transportation and also in smarter living at home also. And because of these implications only nanotechnology is also called as "general purpose technology".

As a "general-purpose technology", it will have multiple uses, not only in commercial field as well as in defence field too that will include making of far more powerful and better weapons and equipments for infantry, air force and navy. Nanotechnology is about building machines at the molecular level. Machines so small they can travel through our blood stream.

Nanotechnology will allow making high-quality products at a very low cost, and also allow making new nanofactories at the same low cost and at a very rapid speed. Nanotechnology offers not just better products, but a vastly improved means of production for e. g. as many copies of data files as we want can be taken out from our computer at a very or no cost.

With time, manufacture of products will become as cheap as the copying of files. So this is what nanotechnology is, and so it is often seen as the next industrial revolution. Nanoscale materials are used in electronic, magnetic and optoelectronic, biomedical, pharmaceutical, cosmetic, energy, catalytic and materials applications.

The word nano we are talking about is not a smallest thing on the planet or in space. The protons, neurons, quarks, leptons and neutrinos are considered as the family of electrons out of which quarks and leptons are the smallest known particles. Protons, neutrons, pions, quarks, are some other sub-atomic particles are smaller than electrons. Nanotechnology is the manufacturing of electronic circuits and mechanical devices by using these particles that means working at molecular level, with takings every particle, every molecule and every atom in concerned.

In this way scientist can prepare a structure and material that will have absolutely new characteristics and function. It is about to emerge as a technology which is a revolutionary, transformative, powerful, and potentially very dangerous or beneficial technology called "The exponential technology".

In fact, there will not be a single industry that will not be changed by nanotechnological applications. Be it a tennis racquets or long-lasting nanoparticle tennis balls. Foot warmers, athlete skin care or a ski wax. Nanotechnology nowadays is progressing towards the delivery system for anti-cancer drugs at the same time research is going on to develop nanofibre which will help create blood vessels, help in treatment of vascular diseases and in heart surgeries.

The purpose of medical devices and nanorobots traveling through the human body is essentially a positive one of searching out and destroying clusters of cancer cells before they spread. Scientists are also working towards the preparation of injectable nanoparticles that will help as medication for treating alcoholism and other related diseases. Because of this it is also called as the future technology.

A lot of money is being invested in this field. In 2004, USA invested more than $400 million into the research area of nanotechnology, facilities, and business development programs and a lots more in the area of publicity is being poured in. On a global scale, these figures multiply exponentially. Even private firms are pumping up a lot amount of money over two billon dollars a year along with the government in the field of nanotechnology.

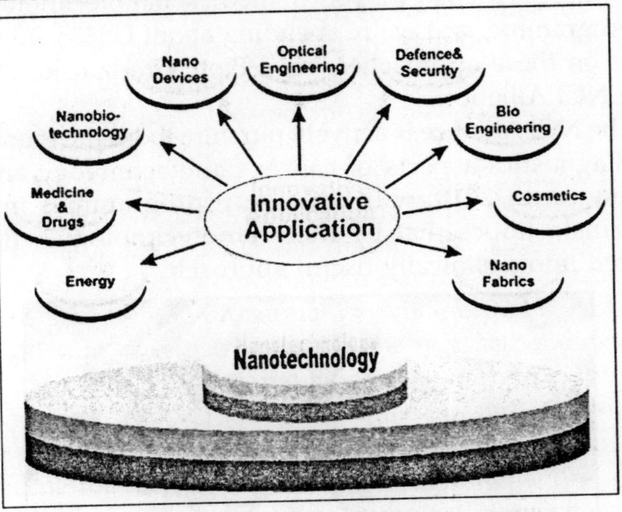

CANCER NANOTECHNOLOGY

With cancer at the forefront of therapeutic applications of nanotechnology, Dan Jones investigates progress in the field.

The US National Cancer Institute (NCI) launched the Alliance for Nanotechnology in Cancer to stimulate and coordinate research in biology, engineering and materials science to push cancer nanotechnology forward. Just over 2 years on, such research is attracting increasing attention: in a round-up of last year's breakthroughs in the burgeoning field of nanotechnology from Forbes magazine, anticancer nanoparticles featured in the top five.

Nanotechnology is being applied to cancer in two broad areas: the development of nanovectors, such as nanoparticles, which can be loaded with drugs or imaging agents and then targeted to tumours, and high-throughput nanosensor devices for detecting the biological signatures of cancer. Combined, such technologies could lead to earlier diagnosis and better treatment for patients with cancer.

Spearheading efforts to expedite the clinical application of these technologies, the NCI currently funds eight Centers of Cancer Nanotechnology Excellence (CCNE) in the United States, in addition to 12 other smaller programmes. "We've pulled most of the key players in medical nanotechnology into this programme, and we're spending about US$35–40 million a year on these approaches,"says Piotr Grodzinski, Director of the NCI Alliance.

The NCI Alliance is actively pursuing both the therapeutic and diagnostics aspects of cancer nanotechnology and will continue until 2010, with follow-up programmes in sight. Grodzinski hopes that by then nanotechnology will have matured into a clinically useful approach.

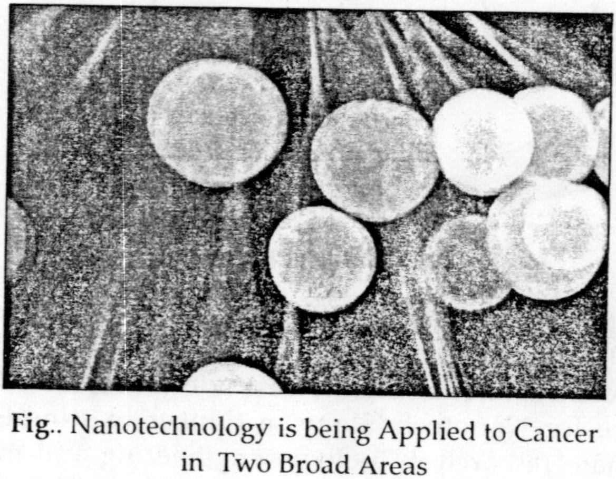

Fig.. Nanotechnology is being Applied to Cancer in Two Broad Areas

Early signs are promising. The research highlighted in the Forbes list was conducted by a team led by Robert Langer, a chemical engineer who is Institute Professor at the Massachusetts Institute of Technology (MIT), USA, and also one of the two principal investigators for the MIT–Harvard CCNE.

Langer, with Omid Farokhzad, Assistant Professor of Anaesthesia at Brigham and Women's Hospital, Boston, USA, and colleagues, uses polymeric nanoparticles coated with aptamers — RNA-based targeting moieties — to guide them towards the tumour, where they bind, enter the cells and then dissolve to spill out their contents — the anticancer drug docetaxel. The nanoparticles are also coated with polyethylene glycol (PEG) to aid their safe passage through the bloodstream and into the tumour cells.

A single injection of such nanoparticles coated with aptamers that bind to prostate-membrane-specific antigen eradicated tumours in a mouse model of prostate cancer. "Extensive animal models show that this approach is both safe and efficacious," says Langer.

One of the key challenges in creating effective nanoparticles is targeting them to appropriate tissues and cells. Although biological targeting using aptamers or antibodies on the surface of nanoparticles is one popular option, other researchers are beginning to exploit the physical characteristics of the particles to guide them to desired locations. "The size, shape, physical properties, density and charge all affect how particles travel through the body, and whether or not they will cross biological membranes," says Mauro Ferrari, a professor of nanotechnology at the University of Texas Health Science Centre, the M. D. Anderson Cancer Centre, and Rice University in Houston, USA.

His work shows that biological barriers such as the vascular wall dominate the distribution of injected nanoparticles in the body, even for particles that are decorated with exquisite biological recognition moieties. A judicious choice of size and shape of a nanovector particle can enhance by orders of magnitude the amount of drug delivered to the target lesion site.

"I believe that the era of 'rational design' of nanoparticles has arrived, and that optimal design will occur based on principles of engineering and physics," Ferrari says. "I believe that the era of 'rational design' of nanoparticles has arrived, and that optimal design will occur based on principles of engineering and physics."

Joseph DeSimone, professor of chemistry and chemical engineering at the University of North Carolina at Chapel Hill, USA, is putting these principles into practice. DeSimone has adapted fabrication technologies from the electronics industry to produce shape-specific organic nanoparticles. "We basically make moulds out of a really low-surface-energy fluoropolymer that allows us to synthesize truly engineered particles with desired characteristics," says DeSimone.

DeSimone's engineered process allows the precise control over particle size (20 nm to >100 μm), particle shape (spheres, cylinders, discs, toroidal), particle composition (organic or inorganic, solid or porous), particle cargo (hydrophilic or hydrophobic therapeutics, biologicals, imaging agents), particle compliance (stiff, deformable) and particle surface properties (Avidin–biotin complexes, targeting peptides, antibodies, aptamers, PEG chains).

Nanoparticles are not the only way to encapsulate a drug or imaging agent into a small carrier, but DeSimone suggests that the nanotechnology approach offers crucial advantages. "With liposomes we kinetically trap what cargo molecule we can, but we can't modify the amount that gets trapped," says DeSimone. "Our moulds enable us to make organic particles loaded with therapeutic cargoes at any amount — 5per cent, 20per cent or whatever we want. "

DeSimone is also taking a cue from naturally occurring particles, such as red blood cells, to produce compliant nanoparticles that can deform to pass across biological barriers such as sinusoids in the spleen or the blood–brain barrier. "The ability to get through these barriers and the flow characteristics of particles — whether they flow through the centre of a capillary or along cell walls — are dictated by size, shape, surface chemistry and compliance," says DeSimone.

Liquidia Technologies Inc., based in Morrisville, North Carolina, USA, was spun-off from DeSimone's laboratory a few years ago to develop this technology platform. Right now, the company is working on feasibility studies, as well as research collaborations with some large pharmaceutical and medical device companies.

"These feasibility studies will run through the rest of 2007, with the goal of focused out-licensing and joint product development deals within the next 18 months," says Luke Roush, Vice President of Business Development at Liquidia.

"These studies will also provide data that will validate application of our technology platform, and help us advance knowledge about how to apply it in areas of clinical need," adds Roush.

The synergy between the therapeutic and diagnostic/ monitoring applications of nanotechnology could be particularly potent. Linda Molnar, a programme officer at the NCI Alliance, sees a future in which new imaging agents, new diagnostic chips and new targeted therapies come together to facilitate a form of personalized medicine in which early and more accurate detection leads to rapid initiation of treatment, followed by diagnostic tests to see whether the patient is responding.

If they do, good; if not, another therapy can be tried and the process repeated — what Molnar refers to as real-time therapeutic monitoring. "The sooner we can detect the cancer and start treatment, and know that we're treating that patient with a therapy that they respond to, the better," says Molnar. "That's why people are excited, and why people at the NCI have made such a large investment in nanotechnology for cancer."

As novel nanomedicine products move from the laboratory to the clinic, the issue of regulatory approval of these new technologies will come to the fore. In many cases these products will combine established drugs with materials already used in FDA-approved therapies, and no special provisions are in place to deal with nanomedicine at present.

An FDA spokesperson said, "All nanotechnology applications will at this point fall within the existing framework for review of all products submitted to CDER, CDER is discussing how nanotech products may be treated differently, if necessary."

Chapter 2

Working Process of Nanotechnology

There's an unprecedented multidisciplinary convergence of scientists dedicated to the study of a world so small, we can't see it — even with a light microscope. That world is the field of nanotechnology, the realm of atoms and nanostructures.

Nanotechnology is so new, no one is really sure what will come of it. Even so, predictions range from the ability to reproduce things like diamonds and food to the world being devoured by self-replicating nanorobots.

In order to understand the unusual world of nanotechnology, we need to get an idea of the units of measure involved. A centimeter is one-hundredth of a meter, a millimeter is one-thousandth of a meter, and a micrometer is one-millionth of a meter, but all of these are still huge compared to the nanoscale. A nanometer (nm) is one-billionth of a meter, smaller than the wavelength of visible light and a hundred-thousandth the width of a human hair.

As small as a nanometer is, it's still large compared to the atomic scale. An atom has a diameter of about 0. 1 nm. An atom's nucleus is much smaller — about 0. 00001 nm. Atoms are the building blocks for all matter in our universe. We and everything around us are made of atoms.

Nature has perfected the science of manufacturing matter molecularly. For instance, our bodies are assembled in a specific manner from millions of living cells. Cells are nature's nanomachines. At the atomic scale, elements are at their most basic level. On the nanoscale, we can potentially put these atoms together to make almost anything.

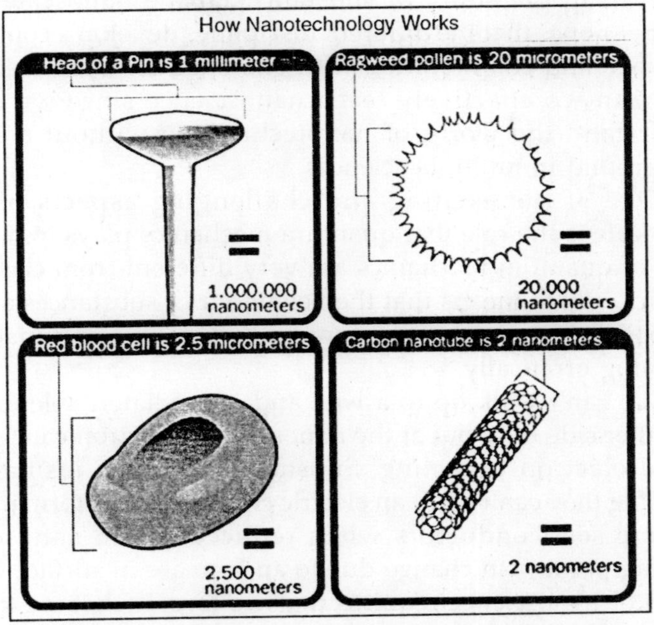

In a lecture called "Small Wonders: The World of Nanoscience," Nobel Prize winner Dr. Horst Störmer said that the nanoscale is more interesting than the atomic scale because the nanoscale is the first point where we can assemble something — it's not until we start putting atoms together that we can make anything useful.

In this article, we'll learn about what nanotechnology means today and what the future of nanotechnology may hold. We'll also look at the potential risks that come with working at the nanoscale.

THE WORLD OF NANOTECHNOLOGY

Experts sometimes disagree about what constitutes the nanoscale, but in general, we can think of nanotechnology dealing with anything measuring between 1 and 100 nm. Larger than that is the microscale, and smaller than that is the atomic scale.

Nanotechnology is rapidly becoming an interdisciplinary field. Biologists, chemists, physicists and engineers are all involved in the study of substances at the nanoscale. Dr. Störmer hopes that the different disciplines develop a common language and communicate with one another. Only then, he says, can we effectively teach nanoscience since we can't understand the world of nanotechnology without a solid background in multiple sciences.

One of the exciting and challenging aspects of the nanoscale is the role that quantum mechanics plays in it. The rules of quantum mechanics are very different from classical physics, which means that the behaviour of substances at the nanoscale can sometimes contradict common sense by behaving erratically.

We can't walk up to a wall and immediately teleport to the other side of it, but at the nanoscale an electron can — it's called electron tunneling. Substances that are insulators, meaning they can't carry an electric charge, in bulk form might become semiconductors when reduced to the nanoscale. Melting points can change due to an increase in surface area. Much of nanoscience requires that we forget what we know and start learning all over again.

So what does this all mean? Right now, it means that scientists are experimenting with substances at the nanoscale to learn about their properties and how we might be able to take advantage of them in various applications. Engineers are trying to use nano-size wires to create smaller, more powerful microprocessors. Doctors are searching for ways to use nanoparticles in medical applications. Still, we've got a long way to go before nanotechnology dominates the technology and medical markets.

NANOWIRES AND CARBON NANOTUBES

Currently, scientists find two nano-size structures of particular interest: nanowires and carbon nanotubes. Nanowires are wires with a very small diameter, sometimes as small as 1 nanometer. Scientists hope to use them to build tiny transistors for computer chips and other electronic devices. In the last couple of years, carbon nanotubes have overshadowed nanowires. We're still learning about these structures, but what we've learned so far is very exciting.

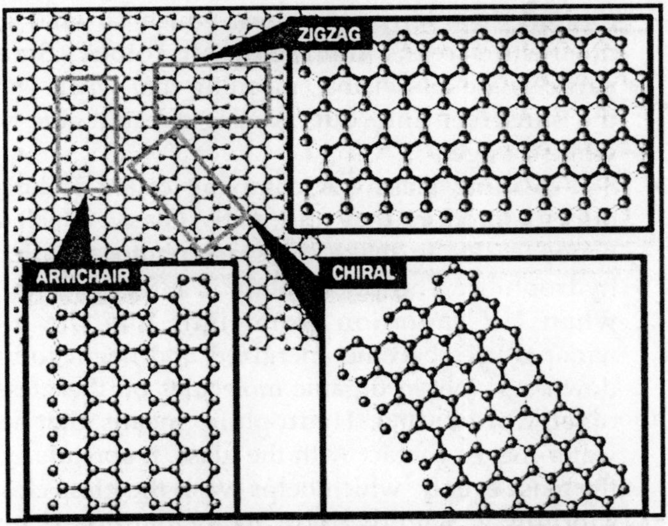

A carbon nanotube is a nano-size cylinder of carbon atoms. Imagine a sheet of carbon atoms, which would look like a sheet of hexagons. If we roll that sheet into a tube, we'd have a carbon nanotube. Carbon nanotube properties depend on how we roll the sheet. In other words, even though all carbon nanotubes are made of carbon, they can be very different from one another based on how we align the individual atoms.

With the right arrangement of atoms, we can create a carbon nanotube that's hundreds of times stronger than steel, but six times lighter. Engineers plan to make building material out of carbon nanotubes, particularly for things like cars and airplanes. Lighter vehicles would mean better fuel efficiency, and the added strength translates to increased passenger safety.

Carbon nanotubes can also be effective semiconductors with the right arrangement of atoms. Scientists are still working on finding ways to make carbon nanotubes a realistic option for transistors in microprocessors and other electronics.

PRODUCTS WITH NANOTECHNOLOGY

- Sunscreen - Many sunscreens contain nanoparticles of zinc oxide or titanium oxide. Older sunscreen formulas use larger particles, which is what gives most sunscreens their whitish colour. Smaller particles are less visible, meaning that when we rub the sunscreen into our skin, it doesn't give we a whitish tinge.
- Self-cleaning glass - A company called Pilkington offers a product they call Activ Glass, which uses nanoparticles to make the glass photocatalytic and hydrophilic. The photocatalytic effect means that when UV radiation from light hits the glass, nanoparticles become energized and begin to break down and loosen organic molecules on the glass (in other words, dirt). Hydrophilic means that when water makes contact with the glass, it spreads across the glass evenly, which helps wash the glass clean.
- Clothing - Scientists are using nanoparticles to enhance our clothing. By coating fabrics with a thin layer of zinc oxide nanoparticles, manufacturers can create clothes that give better protection from UV radiation. Some clothes have nanoparticles in the form of little hairs or whiskers that help repel water and other materials, making the clothing stain-resistant.
- Scratch-resistant coatings - Engineers discovered that adding aluminum silicate nanoparticles to scratch-resistant polymer coatings made the coatings more effective, increasing resistance to chipping and scratching. Scratch-resistant coatings are common on everything from cars to eyeglass lenses.
- Antimicrobial bandages - Scientist Robert Burrell created a process to manufacture antibacterial bandages using nanoparticles of silver. Silver ions block microbes' cellular respiration. In other words, silver smothers harmful cells, killing them.
- Swimming pool cleaners and disinfectants -

EnviroSystems, Inc. developed a mixture (called a nanoemulsion) of nano-sized oil drops mixed with a bactericide. The oil particles adhere to bacteria, making the delivery of the bactericide more efficient and effective.

New products incorporating nanotechnology are coming out every day. Wrinkle-resistant fabrics, deep-penetrating cosmetics, liquid crystal displays (LCD) and other conveniences using nanotechnology are on the market. Before long, we'll see dozens of other products that take advantage of nanotechnology ranging from Intel microprocessors to bio-nanobatteries, capacitors only a few nanometers thick. While this is exciting, it's only the tip of the iceberg as far as how nanotechnology may impact us in the future.

THE FUTURE OF NANOTECHNOLOGY

In the world of "Star Trek," machines called replicators can produce practically any physical object, from weapons to a steaming cup of Earl Grey tea. Long considered to be exclusively the product of science fiction, today some people believe replicators are a very real possibility. They call it molecular manufacturing, and if it ever does become a reality, it could drastically change the world.

Atoms and molecules stick together because they have complementary shapes that lock together, or charges that attract. Just like with magnets, a positively charged atom will stick to a negatively charged atom. As millions of these atoms are pieced together by nanomachines, a specific product will begin to take shape. The goal of molecular manufacturing is to manipulate atoms individually and place them in a pattern to produce a desired structure.

The first step would be to develop nanoscopic machines, called assemblers, that scientists can programme to manipulate atoms and molecules at will. Rice University Professor Richard Smalley points out that it would take a single nanoscopic machine millions of years to assemble a meaningful amount of material.

In order for molecular manufacturing to be practical, we

would need trillions of assemblers working together simultaneously. Eric Drexler believes that assemblers could first replicate themselves, building other assemblers. Each generation would build another, resulting in exponential growth until there are enough assemblers to produce objects.

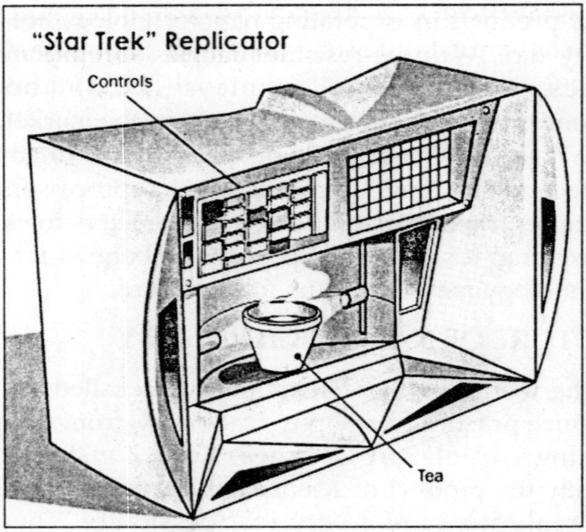

Fig. Star Trek Machine

Trillions of assemblers and replicators could fill an area smaller than a cubic millimeter, and could still be too small for us to see with the naked eye. Assemblers and replicators could work together to automatically construct products, and could eventually replace all traditional labour methods. This could vastly decrease manufacturing costs, thereby making consumer goods plentiful, cheaper and stronger. Eventually, we could be able to replicate anything, including diamonds, water and food. Famine could be eradicated by machines that fabricate foods to feed the hungry.

Nanotechnology may have its biggest impact on the medical industry. Patients will drink fluids containing nanorobots programmed to attack and reconstruct the molecular structure of cancer cells and viruses. There's even speculation that nanorobots could slow or reverse the aging process, and life expectancy could increase significantly. Nanorobots could also be programmed to perform delicate surgeries — such nanosurgeons could work at a level a thousand times more precise than the sharpest scalpel. By working on such a small scale, a nanorobot could operate without leaving the scars that conventional surgery does. Additionally, nanorobots could change our physical appearance. They could be programmed to perform cosmetic surgery, rearranging our atoms to change our ears, nose, eye colour or any other physical feature we wish to alter.

Nanotechnology has the potential to have a positive effect on the environment. For instance, scientists could programme airborne nanorobots to rebuild the thinning ozone layer. Nanorobots could remove contaminants from water sources and clean up oil spills. Manufacturing materials using the bottom-up method of nanotechnology also creates less pollution than conventional manufacturing processes. Our dependence on non-renewable resources would diminish with nanotechnology. Cutting down trees, mining coal or drilling for oil may no longer be necessary — nanomachines could produce those resources.

Many nanotechnology experts feel that these applications are well outside the realm of possibility, at least for the foreseeable future. They caution that the more exotic applications are only theoretical. Some worry that nanotechnology will end up like virtual reality — in other words, the hype surrounding nanotechnology will continue to build until the limitations of the field become public knowledge, and then interest (and funding) will quickly dissipate.

CHALLENGES AND RISKS OF NANOTECHNOLOGY

The most immediate challenge in nanotechnology is that we need to learn more about materials and their properties at the nanoscale. Universities and corporations across the world are rigorously studying how atoms fit together to form larger structures. We're still learning about how quantum mechanics impact substances at the nanoscale.

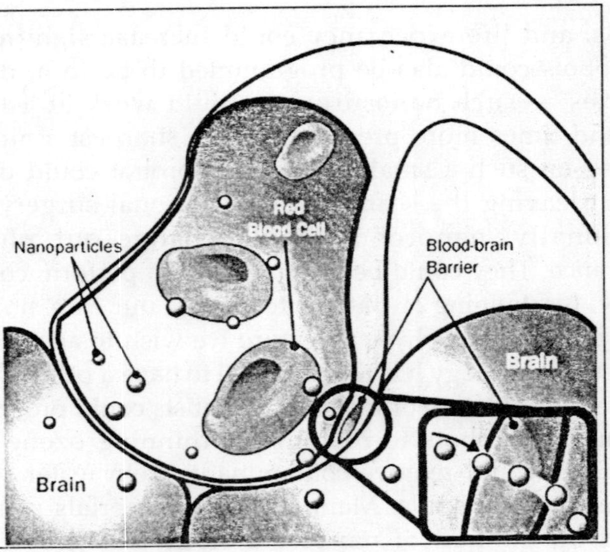

Fig. Blood-BrainBarrier

Because elements at the nanoscale behave differently than they do in their bulk form, there's a concern that some nanoparticles could be toxic. Some doctors worry that the nanoparticles are so small, that they could easily cross the blood-brain barrier, a membrane that protects the brain from harmful chemicals in the bloodstream. If we plan on using nanoparticles to coat everything from our clothing to our highways, we need to be sure that they won't poison us.

Closely related to the knowledge barrier is the technical barrier. In order for the incredible predictions regarding nanotechnology to come true, we have to find ways to mass produce nano-size products like transistors and nanowires. While we can use nanoparticles to build things like tennis rackets and make wrinkle-free fabrics, we can't make really complex microprocessor chips with nanowires yet.

HOW ATOMS WORK

It has been said that during the 20th century, man harnessed the power of the atom. We made atomic bombs and generated electricity by nuclear power. We even split the atom into smaller pieces called subatomic particles. But what exactly is an atom? What is it made of? What does it look like? The pursuit of the structure of the atom has married many areas of chemistry and physics in perhaps one of the greatest contributions of modern science!

In this article, we will follow this fascinating story of how discoveries in various fields of science resulted in our modern view of the atom. We will look at the consequences of knowing the atom's structure and how this structure will lead to new technologies.

WHAT IS AN ATOM

The modern view of an atom has come from many fields of chemistry and physics. The idea of an atom came from ancient Greek science/philosophy and from the results of 18th and 19th century chemistry:

- Concept of the atom
- Measurements of atomic mass
- Repeating or periodic relationship between the elements

Concept of the Atom

Dalton's atomic theory formed the groundwork of chemistry at that time. Dalton envisioned atoms as tiny spheres with hooks on them. With these hooks, one atom could combine with another in definite proportions. But some elements could combine to make different compounds (e. g., hydrogen + oxygen could make water or hydrogen peroxide).

So, he could not say anything about the numbers of each atom in the molecules of specific substances. Did water have one oxygen with one hydrogen or one oxygen with two hydrogens? This point was resolved when chemists figured out how to weigh atoms.

Important Terms about atomic Chemistry:

- Atom - smallest piece of an element that keeps its chemical properties
- Compound - substance that can be broken into elements by chemical reactions

- Electron - particle orbiting the nucleus of an atom with a negative charge (mass = 9.10×10^{-28} grams)
- Element - substance that cannot be broken down by chemical reactions
- Ion - electrically charged atom (i. e., excess positive or negative charge)
- Molecule - smallest piece of a compound that keeps its chemical properties (made of two or more atoms)
- Neutron - particle in the nucleus of an atom with no charge (mass = 1.675×10^{-24} grams)
- Nucleus - dense, central core of an atom (made of protons and neutrons)
- Proton - particle in the nucleus of an atom with a positive charge (mass = 1.673×10^{-24} grams)

WEIGHTS OF ATOM

The ability to weigh atoms came about by an observation from an Italian chemist named Amadeo Avogadro. Avogadro was working with gases (nitrogen, hydrogen, oxygen, chlorine) and noticed that when temperature and pressure was the same, these gases combined in definite volume ratios. For example:

- One liter of nitrogen combined with three liters of hydrogen to form ammonia (NH_3)
- One liter of hydrogen combined with one liter of chlorine to make hydrogen chloride (HCl)

Avogadro said that at the same temperature and pressure, equal volumes of the gases had the same number of molecules. So, by weighing the volumes of gases, he could determine the ratios of atomic masses.

For example, a liter of oxygen weighed 16 times more than a liter of hydrogen, so an atom of oxygen must be 16 times the mass of an atom of hydrogen. Work of this type resulted in a relative mass scale for elements in which all of the elements related to carbon.

Once the relative mass scale was made, later experiments were able to relate the mass in grams of a substance to the number of atoms and an atomic mass unit (amu) was found; 1 amu or Dalton is equal to 1.66×10^{-24} grams.

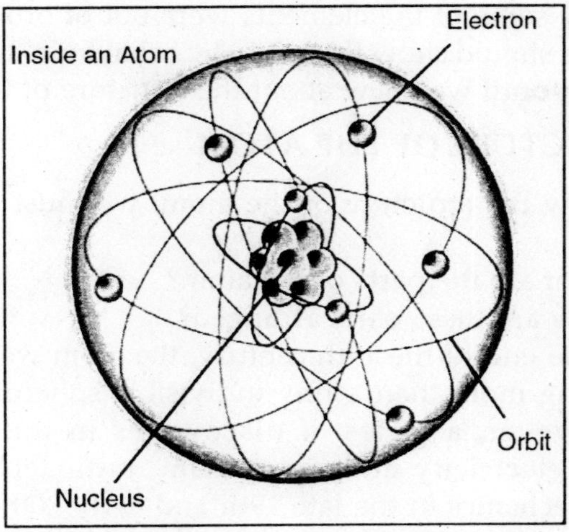

Fig. Simplest model of an Atom

At this time, chemists knew the atomic masses of elements and their chemical properties, and an astonishing phenomenon jumped out at them!

PROPERTIES OF ELEMENTS

At the time that atomic masses had been discovered, a Russian chemist named Dimitri Mendeleev was writing a textbook. For his book, he began to organize elements in terms of their properties by placing the elements and their newly discovered atomic masses in cards.

He arranged the elements by increasing atomic mass and noticed that elements with similar properties appeared at regular intervals or periods. Mendeleev's table had two problems:

- There were some gaps in his "periodic table. "
- When grouped by properties, most elements had increasing atomic masses, but some were out of order.

To explain the gaps, Mendeleev said that the gaps were due to undiscovered elements. In fact, his table successfully predicted the existence of gallium and germanium, which were discovered later. However, Mendeleev was never able to explain why some of the elements were out of order or why the elements should show this periodic behaviour. This would have to wait until we knew about the structure of the atom.

THE STRUCTURE OF THE ATOM

To know the structure of the atom, we must know the following:

- What are the parts of the atom?
- How are these parts arranged?

Near the end of the 19th century, the atom was thought to be nothing more than a tiny indivisible sphere (Dalton's view). However, a series of discoveries in the fields of chemistry, electricity and magnetism, radioactivity, and quantum mechanics in the late 19th and early 20th centuries changed all of that. Here is what these fields contributed:

- The parts of the atom:
  - chemistry and electromagnetism —> electron (first subatomic particle)
  - radioactivity —> nucleus
    a. proton
    b. neutron
- How the atom is arranged - quantum mechanics puts it all together:
  - atomic spectra —> Bohr model of the atom
  - wave-particle duality —> Quantum model of the atom

Discovering the Electron

In the late 19th century, chemists and physicists were studying the relationship between electricity and matter. They were placing high voltage electric currents through glass tubes filled with low-pressure gas (mercury, neon, xenon) much like neon lights. Electric current was carried from one electrode (cathode) through the gas to the other electrode (anode) by a beam called cathode rays. In 1897, a British physicist, J. J. Thomson did a series of experiments with the following results:

- He found that if the tube was placed within an electric or magnetic field, then the cathode rays could be deflected or moved (this is how the the cathode ray tube (CRT) on our television works).

- By applying an electric field alone, a magnetic field alone, or both in combination, Thomson could measure the ratio of the electric charge to the mass of the cathode rays.
- He found the same charge to mass ratio of cathode rays was seen regardless of what material was inside the tube or what the cathode was made of.

Thomson concluded the following:

- Cathode rays were made of tiny, negatively charged particles, which he called electrons.
- The electrons had to come from inside the atoms of the gas or metal electrode.
- Because the charge to mass ratio was the same for any substance, the electrons were a basic part of all atoms.
- Because the charge to mass ratio of the electron was very high, the electron must be very small.

Later, an American Physicist named Robert Milikan measured the electrical charge of an electron. With these two numbers (charge, charge to mass ratio), physicists calculated the mass of the electron as 9. 10 x 10^{-28} grams. For comparison, a U. S. penny has a mass of 2. 5 grams; so, 2. 7 x 10^{27} or 2. 7 billion billion billion electrons would weigh as much as a penny!

Two other conclusions came from the discovery of the electron:

- Because the electron was negatively charged and atoms are electrically neutral, there must be a positive charge somewhere in the atom.
- Because electrons are so much smaller than atoms, there must be other, more massive particles in the atom.

From these results, Thomson proposed a model of the atom that was like a watermelon. The red part was the positive charge and the seeds were the electrons.

RADIOACTIVITY

About the same time as Thomson's experiments with

cathode rays, physicists such as by Henri Becquerel, Marie Curie, Pierre Curie, and Ernest Rutherford were studying radioactivity. Radioactivity was characterized by three types of emitted rays:

- Alpha particles - positively charged and massive. Ernest Rutherford showed that these particles were the nucleus of a helium atom.
- Beta particles - negatively charged and light (later shown to be electrons).
- Gamma rays - neutrally charged and no mass (i. e., energy).

The experiment from radioactivity that contributed most to our knowledge of the structure of the atom was done by Rutherford and his colleagues. Rutherford bombarded a thin foil of gold with a beam of alpha particles and looked at the beams on a fluorescent screen, he noticed the following:

- Most of the particles went straight through the foil and struck the screen.
- Some (0. 1 per cent) were deflected or scattered in front (at various angles) of the foil, while others were scattered behind the foil.

Rutherford concluded that the gold atoms were mostly empty space, which allowed most of the alpha particles through. However, some small region of the atom must have been dense enough to deflect or scatter the alpha particle. He called this dense region the nucleus; the nucleus comprised most of the mass of the atom.

Later, when Rutherford bombarded nitrogen with alpha particles, a positively charged particle that was lighter than the alpha particle was emitted. He called these particles protons and realized that they were a fundamental particle in the nucleus. Protons have a mass of 1.673×10^{-24} grams, about 1,835 times larger than an electron!

However, protons could not be the only particle in the nucleus because the number of protons in any given element (determined by the electrical charge) was less than the weight of the nucleus. Therefore, a third, neutrally charged particle must exist! It was James Chadwick, a British physicist and coworker of Rutherford, who discovered the third subatomic particle, the neutron.

Chadwick bombarded beryllium foil with alpha particles and noticed a neutral radiation coming out. This neutral radiation could in turn knock protons out of the nuclei of other substances. Chadwick concluded that this radiation was a stream of neutrally charged particles with about the same mass as a proton; the neutron has a mass of 1. 675 x 10^{-24} grams.

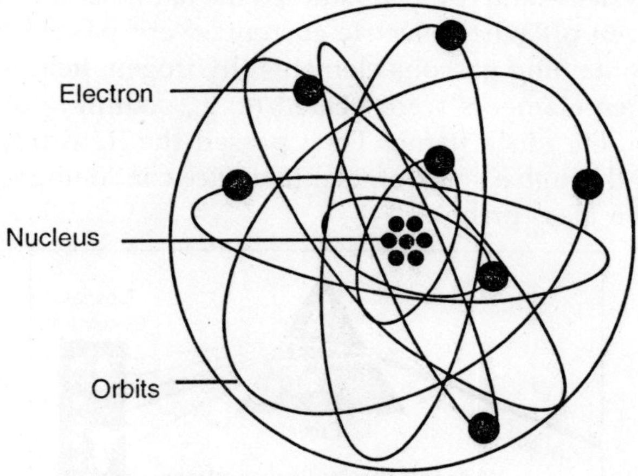

Fig. Rutherford's view of the atom

Now that the parts of the atom were known, how were they arranged to make an atom? Rutherford's gold foil experiment indicated that the nucleus was in the centre of the atom and that the atom was mostly empty space. So, he envisioned the atom as the positively charged nucleus in the centre with the negatively charged electrons circling around it much like a planet with moons.

Although he had no evidence that the electrons circled the nucleus, his model seemed reasonable; however, it presented a problem. As the electrons moved in a circle, they would lose energy and give off light. The loss of energy would slow the electrons down. Like any satellite, the slowing electrons would fall into the nucleus. In fact, it was calculated that a Rutherford atom would last only billionths of a second before collapsing! Something was missing!

QUANTUM MECHANICS

At the same time that discoveries were being made with radioactivity, physicists and chemists were studying how light interacted with matter. These studies began the field of quantum mechanics and helped solve the structure of the atom.

Quantum Mechanics Sheds Light on the Atom: The Bohr Model

Physicists and chemists studied the nature of the light that was given off when electric currents were passed through tubes containing gaseous elements (hydrogen, helium, neon) and when elements were heated (e. g., sodium, potassium, calcium, etc.) in a flame. They passed the light from these sources through a spectrometer (a device containing a narrow slit and a glass prism).

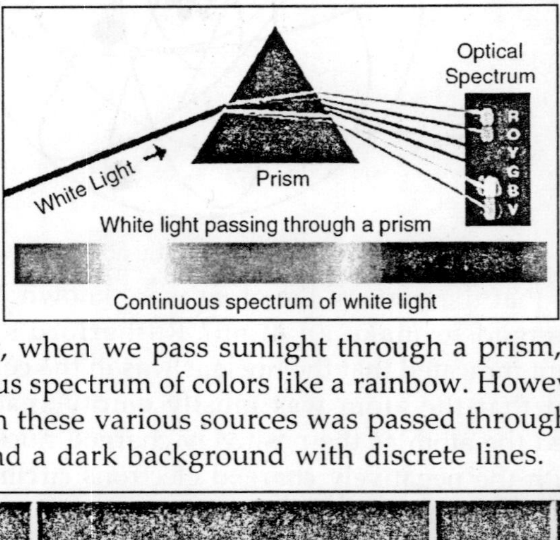

Now, when we pass sunlight through a prism, we get a continuous spectrum of colors like a rainbow. However, when light from these various sources was passed through a prism, they found a dark background with discrete lines.

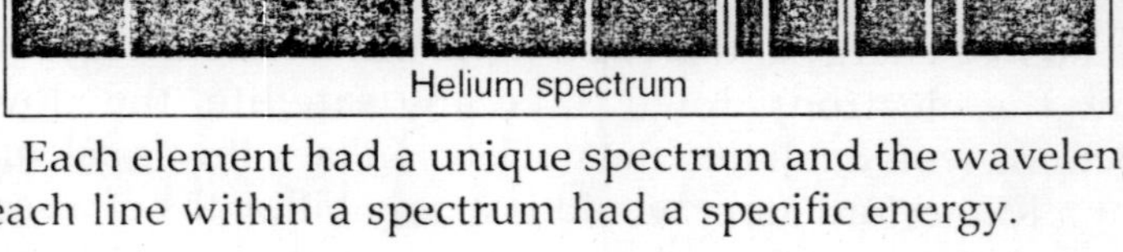

Each element had a unique spectrum and the wavelength of each line within a spectrum had a specific energy.

In 1913, a Danish physicist named Niels Bohr put Rutherford's findings together with the observed spectra to come up with a new model of the atom in a real leap of intuition. Bohr suggested that the electrons orbiting an atom could only exist at certain energy levels (i. e., distances) from the nucleus, not at continuous levels as might be expected from Rutherford's model.

When atoms in the gas tubes absorbed the energy from the electric current, the electrons became excited and jumped from low energy levels (close to the nucleus) to high energy levels (farther out from the nucleus).

The excited electrons would fall back to their original levels and emit energy as light. Because there were specific differences between the energy levels, only specific wavelengths of light were seen in the spectrum (i. e., lines).

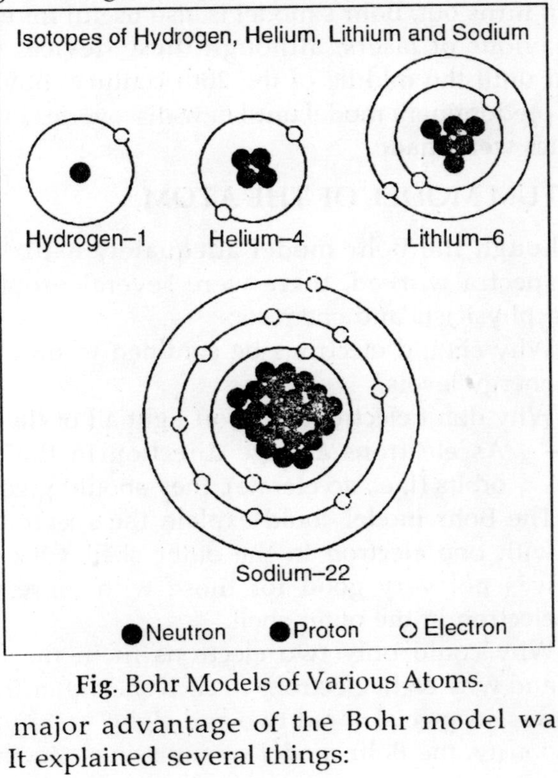

Fig. Bohr Models of Various Atoms.

The major advantage of the Bohr model was that it worked. It explained several things:

- Atomic spectra
- Periodic behaviour of elements - elements with similar properties had similar atomic spectra.
  - Each electron orbit of the same size or energy (shell) could only hold so many electrons.
- First shell = two electrons
- Second shell = eight electrons
- Third shell and higher = eight electrons
  - When one shell was filled, electrons were found at higher levels.
  - Chemical properties were based on the number of electrons in the outermost shell.
- Elements with full outer shells do not react.
- Other elements take or give up electrons to get a full outer shell.

As it turns out, Bohr's model is also useful for explaining the behaviour of lasers, although these devices were not invented until the middle of the 20th century. Bohr's model was the predominant model until new discoveries in quantum mechanics were made.

QUANTUM MODEL OF THE ATOM

Although the Bohr model adequately explained how atomic spectra worked, there were several problems that bothered physicists and chemists:

- Why should electrons be confined to only specified energy levels?
- Why don't electrons give off light all of the time?
  - As electrons change direction in their circular orbits (i. e., accelerate), they should give off light.
- The Bohr model could explain the spectra of atoms with one electron in the outer shell very well, but was not very good for those with more than one electron in the outer shell.
- Why could only two electrons fit in the first shell and why eight electrons in each shell after that? What was so special about two and eight?

Obviously, the Bohr model was missing something!

In 1924, a French physicist named Louis de Broglie suggested that, like light, electrons could act as both particles and waves. De Broglie's hypothesis was soon confirmed in experiments that showed electron beams could be diffracted or bent as they passed through a slit much like light could. So, the waves produced by an electron confined in its orbit about the nucleus sets up a standing wave of specific wavelength, energy and frequency (i. e., Bohr's energy levels) much like a guitar string sets up a standing wave when plucked.

Another question quickly followed de Broglie's idea. If an electron traveled as a wave, could we locate the precise position of the electron within the wave? A German physicist, Werner Heisenberg, answered no in what he called the uncertainty principle:

- To view an electron in its orbit, we must shine a wavelength of light on it that is smaller than the electron's wavelength.
- This small wavelength of light has a high energy.
- The electron will absorb that energy.
- The absorbed energy will change the electron's position.

We can never know both the momentum and position of an electron in an atom. Therefore, Heisenberg said that we shouldn't view electrons as moving in well-defined orbits about the nucleus!

With de Broglie's hypothesis and Heisenberg's uncertainty principle in mind, an Austrian physicist named Erwin Schrodinger derived a set of equations or wave functions in 1926 for electrons. Electrons confined in their orbits would set up standing waves and we could describe only the probability of where an electron could be. The distributions of these probabilities formed regions of space about the nucleus were called orbitals. Orbitals could be described as electron density clouds. The densest area of the cloud is where we have the greatest probability of finding the electron and the least dense area is where we have the lowest probability of finding the electron.

WAVE FUNCTIONS

The wave function of each electron can be described as a set of three quantum numbers:

- Principal number (n) - describes the energy level.
- Altazimuth number (l) - how fast the electron moves in its orbit (angular momentum); like how fast a CD spins (rpm). This is related to the shape of the orbital.
- Magnetic (m) - its orientation in space.

It was later suggested that no two electrons could be in the exact same state, so a fourth quantum number was added. This number was related to the direction that the electron spins while it is moving in its orbit (i. e., clockwise, counterclockwise). Only two electrons could share the same orbital, one spinning clockwise and the other spinning counterclockwise.

The orbitals had different shapes and maximum numbers at any level:

- S (sharp) - spherical (max = 1)
- P (principal) - dumb-bell shaped (max = 3)
- D (diffuse) - four-lobe-shaped (max = 5)
- F (fundamental) - six-lobe shaped (max = 7)

The names of the orbitals came from names of atomic spectral features before quantum mechanics was formally invented. Each orbital can hold only two electrons. Also, the orbitals have a specific order of filling, generally:

- S
- P
- D
- F

However, there is some overlap.

The resulting model of the atom is called the quantum model of the atom.

Sodium has 11 electrons distributed in the following energy levels:

- One s orbital - two electrons
- One s orbital - two electrons and three p orbitals (two electrons each)
- One s orbital - one electron

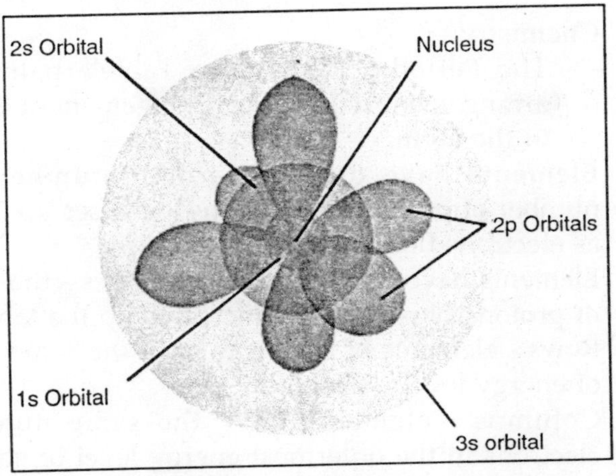

Fig. Quantum Model of a Sodium Atom.

Right now, the quantum model is the most realistic vision of the overall structure of the atom. It explains much of what we know about chemistry and physics. Here are some examples:

Table: The modern periodic table of the elements (elements are ordered based on atomic number rather than mass).

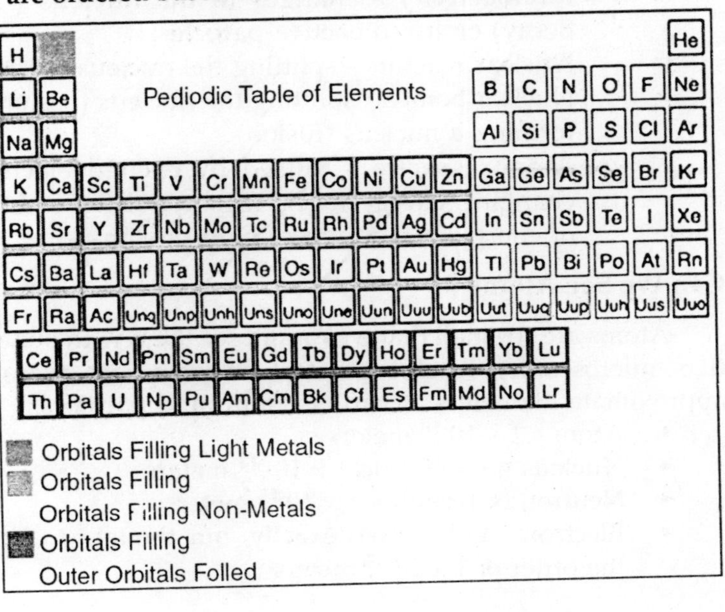

- Chemistry:
  - The Periodic Table - the Table's pattern and arrangement reflects the arrangement of electrons in the atom.
- Elements have different atomic numbers - the number of protons or electrons increases up the table as electrons fill the shells.
- Elements have different atomic masses - the number of protons plus neutrons increases up the table.
- Rows - elements of each row have the same number of energy levels (shells).
- Columns - elements have the same number of electrons in the outermost energy level or shell (one to eight).
  - Chemical reactions - exchange of electrons between various atoms (giving, taking, or sharing). Exchange involves electrons in the outermost energy level in attempts to fill the outermost shell (i. e., most stable form of the atom).
- Physics
  - Radioactivity - changes in the nucleus (i. e., decay) emit radioactive particles.
  - Nuclear reactors - splitting the nucleus (fission)
  - Nuclear bombs - splitting the nucleus (fission) or forming a nucleus (fusion)
  - Atomic spectra - caused by excited electrons changing energy levels (absorption or emission of energy in the form of light photons).

Can We See Atoms

Atoms are so small that we cannot see them with our eyes (i. e., microscopic). To give us a feel for some sizes, these are approximate diameters of various atoms and particles:

- Atom = 1×10^{-10} meters
- Nucleus = 1×10^{-15} to 1×10^{-14} meters
- Neutron or proton = 1×10^{-15} meters
- Electron - not known exactly, but thought to be on the order of 1×10^{-18} meters

We cannot see an atom with a light microscope. However, in 1981, a type of microscope called a scanning tunneling microscope (STM) was developed. The STM consists of the following:

- A very small, sharp tip that conducts electricity (probe)
- A rapid piezoelectric scanning device to which the tip is mounted
- Electronic components to supply current to the tip, control the scanner and accept the signals from the motion sensor
- Computer to control the system and do data analysis (data collection, processing, display)

The STM works like this:

- A current is supplied to the tip (probe) while the scanner rapidly moves the tip across the surface of a conducting sample.
- When the tip encounters an atom, the flow of electrons between the atom and the tip changes.
- The computer registers the change in current with the x,y-position of the atom.
- The scanner continues to position the tip over each x,y-point on the sample surface, registering a current for each point.
- The computer collects the data and plots a map of current over the surface that corresponds to a map of the atomic positions.

The process is much like an old phonograph where the needle is the tip and the grooves in the vinyl record are the atoms. The STM tip moves over the atomic contour of the surface, using tunneling current as a sensitive detector of atomic position.

The STM and new variations of this microscope allow us to see atoms. In addition, the STM can be used to manipulate atoms as shown here:

Atoms can be moved and molded to make various devices such as molecular motors. In summary, science in the 20th century has revealed the structure of the atom. Scientists are now conducting experiments to reveal details of the structure of the nucleus and the forces that hold it together.

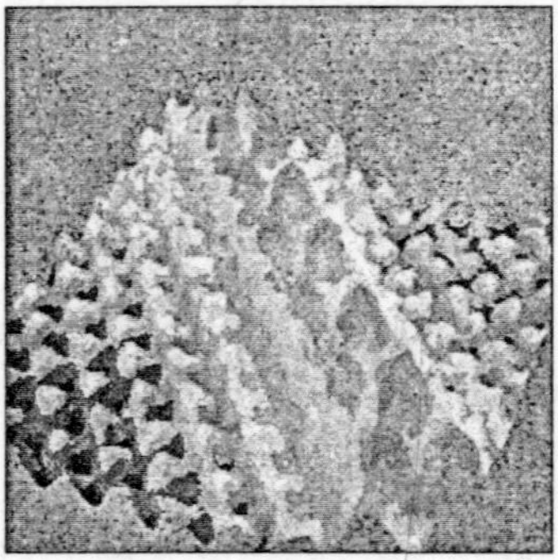

Fig. STM image (7 nm x 7 nm) of a Single Zigzag Chain of Cesium Atoms (red) on a Gallium-Arsenside Surface (blue)

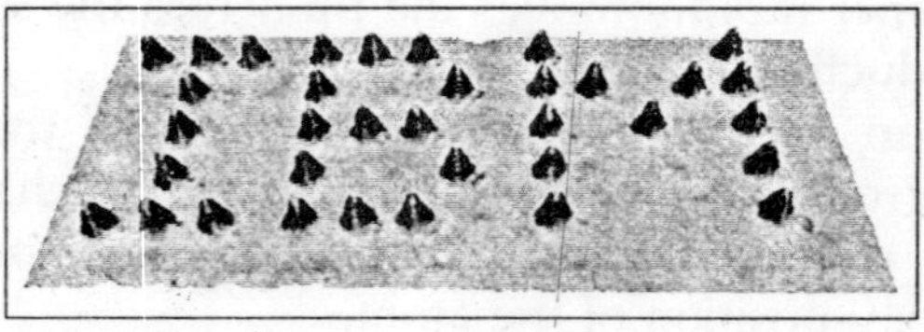

Fig. Atoms can be positioned on a surface using the STM tip, creating a custom pattern on the surface.

HOW MICROPROCESSORS WORK

The computer we are using to read this page uses a microprocessor to do its work. The microprocessor is the heart of any normal computer, whether it is a desktop machine, a server or a laptop. The microprocessor we are using might be a Pentium, a K6, a PowerPC, a Sparc or any of the many other brands and types of microprocessors, but they all do approximately the same thing in approximately the same way.

A microprocessor — also known as a CPU or central processing unit — is a complete computation engine that is fabricated on a single chip. The first microprocessor was the Intel 4004, introduced in 1971. The 4004 was not very powerful — all it could do was add and subtract, and it could only do that 4 bits at a time.

But it was amazing that everything was on one chip. Prior to the 4004, engineers built computers either from collections of chips or from discrete components (transistors wired one at a time). The 4004 powered one of the first portable electronic calculators.

If we have ever wondered what the microprocessor in our computer is doing, or if we have ever wondered about the differences between types of microprocessors, then read on. In this article, we will learn how fairly simple digital logic techniques allow a computer to do its job, whether its playing a game or spell checking a document!

MICROPROCESSOR PROGRESSION: INTEL

The first microprocessor to make it into a home computer was the Intel 8080, a complete 8-bit computer on one chip, introduced in 1974. The first microprocessor to make a real splash in the market was the Intel 8088, introduced in 1979 and incorporated into the IBM PC (which first appeared around 1982).

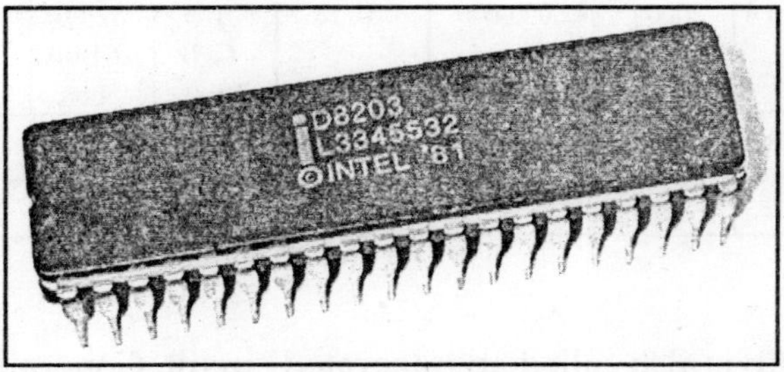

Fig. The Intel 8080 was the First Microprocessor in a Home Computer.

If we are familiar with the PC market and its history, we know that the PC market moved from the 8088 to the 80286 to the 80386 to the 80486 to the Pentium to the Pentium II to the Pentium III to the Pentium 4. All of these microprocessors are made by Intel and all of them are improvements on the basic design of the 8088.

The Pentium 4 can execute any piece of code that ran on the original 8088, but it does it about 5,000 times faster! The following table helps we to understand the differences between the different processors that Intel has introduced over the years.

Name	Date	Transistors	Microns	Clock speed	Data width	MIPS
8080	1974	6,000	6	2 MHz	8 bits	0. 64
8088	1979	29,000	3	5 MHz	16 bits 8-bit bus	0. 33
80286	1982	134,000	1. 5	6 MHz	16 bits	1
80386	1985	275,000	1. 5	16 MHz	32 bits	5
80486	1989	1,200,000	1	25 MHz	32 bits	20
Pentium	1993	3,100,000	0. 8	60 MHz	32 bits 64-bit bus	100
Pentium II	1997	7,500,000	0. 35	233 MHz	32 bits 64-bit bus	~300
Pentium III	1999	9,500,000	0. 25	450 MHz	32 bits 64-bit bus	~510
Pentium 4	2000	42,000,000	0. 18	1. 5 GHz	32 bits 64-bit bus	~1,700
Pentium 4 "Prescott"	2004	125,000,000	0. 09	3. 6 GHz	32 bits 64-bit bus	~7,000

Chip

A chip is also called an integrated circuit. Generally it is a small, thin piece of silicon onto which the transistors making up the microprocessor have been etched. A chip might be as large as an inch on a side and can contain tens of millions of transistors. Simpler processors might consist of a few thousand transistors etched onto a chip just a few millimeters square.

- The date is the year that the processor was first introduced. Many processors are re-introduced at higher clock speeds for many years after the original release date.
- Transistors is the number of transistors on the chip.

We can see that the number of transistors on a single chip has risen steadily over the years.

- Microns is the width, in microns, of the smallest wire on the chip. For comparison, a human hair is 100 microns thick. As the feature size on the chip goes down, the number of transistors rises.
- Clock speed is the maximum rate that the chip can be clocked at.
- Data Width is the width of the ALU. An 8-bit ALU can add/subtract/multiply/etc. two 8-bit numbers, while a 32-bit ALU can manipulate 32-bit numbers. An 8-bit ALU would have to execute four instructions to add two 32-bit numbers, while a 32-bit ALU can do it in one instruction. In many cases, the external data bus is the same width as the ALU, but not always. The 8088 had a 16-bit ALU and an 8-bit bus, while the modern Pentiums fetch data 64 bits at a time for their 32-bit ALUs.
- MIPS stands for "millions of instructions per second" and is a rough measure of the performance of a CPU. Modern CPUs can do so many different things that MIPS ratings lose a lot of their meaning, but we can get a general sense of the relative power of the CPUs from this column.

From this table we can see that, in general, there is a relationship between clock speed and MIPS. The maximum clock speed is a function of the manufacturing process and delays within the chip.

There is also a relationship between the number of transistors and MIPS. For example, the 8088 clocked at 5 MHz but only executed at 0. 33 MIPS (about one instruction per 15 clock cycles). Modern processors can often execute at a rate of two instructions per clock cycle.

MICROPROCESSOR LOGIC

To understand how a microprocessor works, it is helpful to look inside and learn about the logic used to create one. In the process we can also learn about assembly language — the native language of a microprocessor — and many of the things that engineers can do to boost the speed of a processor.

Fig. Intel Pentium 4 Processor

A microprocessor executes a collection of machine instructions that tell the processor what to do. Based on the instructions, a microprocessor does three basic things:

- Using its ALU (Arithmetic/Logic Unit), a microprocessor can perform mathematical operations like addition, subtraction, multiplication and division. Modern microprocessors contain complete floating point processors that can perform extremely sophisticated operations on large floating point numbers.
- A microprocessor can move data from one memory location to another.
- A microprocessor can make decisions and jump to a new set of instructions based on those decisions.

There may be very sophisticated things that a microprocessor does, but those are its three basic activities. The following diagram shows an extremely simple microprocessor capable of doing those three things:

This is about as simple as a microprocessor gets. This microprocessor has:

- An address bus (that may be 8, 16 or 32 bits wide) that sends an address to memory
- A data bus (that may be 8, 16 or 32 bits wide) that can send data to memory or receive data from memory

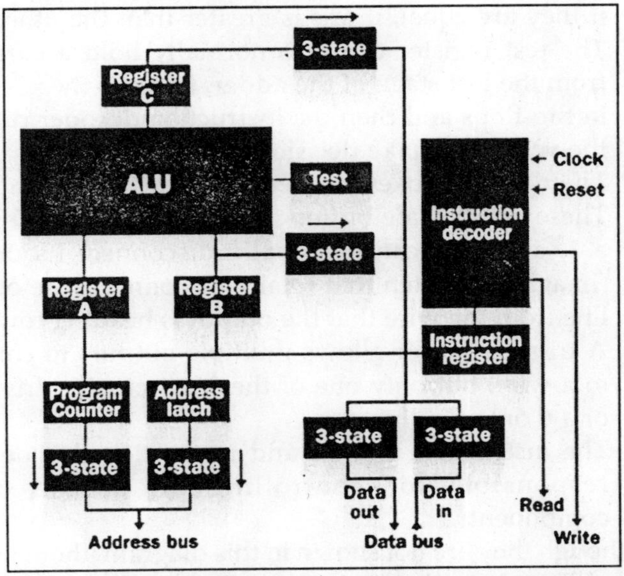

- An RD (read) and WR (write) line to tell the memory whether it wants to set or get the addressed location
- A clock line that lets a clock pulse sequence the processor
- A reset line that resets the programme counter to zero (or whatever) and restarts execution

Let's assume that both the address and data buses are 8 bits wide in this example.

Here are the components of this simple microprocessor:

- Registers A, B and C are simply latches made out of flip-flops.
- The address latch is just like registers A, B and C.
- The programme counter is a latch with the extra ability to increment by 1 when told to do so, and also to reset to zero when told to do so.
- The ALU could be as simple as an 8-bit adder, or it might be able to add, subtract, multiply and divide 8-bit values. Let's assume the latter here.
- The test register is a special latch that can hold values from comparisons performed in the ALU. An ALU can normally compare two numbers and determine

if they are equal, if one is greater than the other, etc. The test register can also normally hold a carry bit from the last stage of the adder. It stores these values in flip-flops and then the instruction decoder can use the values to make decisions.

- There are six boxes marked "3-State" in the diagram. These are tri-state buffers. A tri-state buffer can pass a 1, a 0 or it can essentially disconnect its output (imagine a switch that totally disconnects the output line from the wire that the output is heading toward). A tri-state buffer allows multiple outputs to connect to a wire, but only one of them to actually drive a 1 or a 0 onto the line.
- The instruction register and instruction decoder are responsible for controlling all of the other components.

Although they are not shown in this diagram, there would be control lines from the instruction decoder that would:

- Tell the A register to latch the value currently on the data bus
- Tell the B register to latch the value currently on the data bus
- Tell the C register to latch the value currently output by the ALU
- Tell the programme counter register to latch the value currently on the data bus
- Tell the address register to latch the value currently on the data bus
- Tell the instruction register to latch the value currently on the data bus
- Tell the programme counter to increment
- Tell the programme counter to reset to zero
- Activate any of the six tri-state buffers (six separate lines)
- Tell the ALU what operation to perform
- Tell the test register to latch the ALU's test bits
- Activate the RD line
- Activate the WR line

Coming into the instruction decoder are the bits from the test register and the clock line, as well as the bits from the instruction register.

MICROPROCESSOR MEMORY

The address and data buses, as well as the RD and WR lines. These buses and lines connect either to RAM or ROM — generally both. In our sample microprocessor, we have an address bus 8 bits wide and a data bus 8 bits wide. That means that the microprocessor can address (2^8) 256 bytes of memory, and it can read or write 8 bits of the memory at a time. Let's assume that this simple microprocessor has 128 bytes of ROM starting at address 0 and 128 bytes of RAM starting at address 128.

Fig. ROM chip

ROM stands for read-only memory. A ROM chip is programmed with a permanent collection of pre-set bytes. The address bus tells the ROM chip which byte to get and place on the data bus. When the RD line changes state, the ROM chip presents the selected byte onto the data bus.

RAM stands for random-access memory. RAM contains bytes of information, and the microprocessor can read or write to those bytes depending on whether the RD or WR line is signaled. One problem with today's RAM chips is that they forget everything once the power goes off. That is why the computer needs ROM.

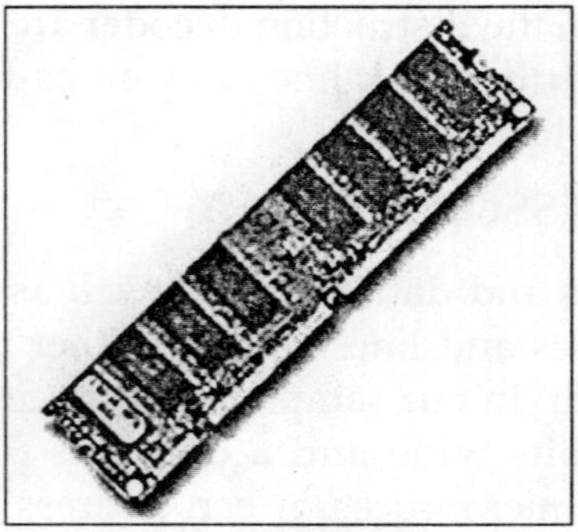

Fig. RAM Chip

By the way, nearly all computers contain some amount of ROM (it is possible to create a simple computer that contains no RAM — many microcontrollers do this by placing a handful of RAM bytes on the processor chip itself — but generally impossible to create one that contains no ROM). On a PC, the ROM is called the BIOS (Basic Input/Output System).

When the microprocessor starts, it begins executing instructions it finds in the BIOS. The BIOS instructions do things like test the hardware in the machine, and then it goes to the hard disk to fetch the boot sector.

This boot sector is another small programme, and the BIOS stores it in RAM after reading it off the disk. The microprocessor then begins executing the boot sector's instructions from RAM. The boot sector programme will tell the microprocessor to fetch something else from the hard disk into RAM, which the microprocessor then executes, and so on. This is how the microprocessor loads and executes the entire operating system.

MICROPROCESSOR INSTRUCTIONS

Even the incredibly simple microprocessor will have a fairly large set of instructions that it can perform. The collection of instructions is implemented as bit patterns, each one of which has a different meaning when loaded into the instruction register. Humans are not particularly good at remembering bit patterns, so a set of short words are defined to represent the different bit patterns. This collection of words is called the assembly language of the processor. An assembler can translate the words into their bit patterns very easily, and then the output of the assembler is placed in memory for the microprocessor to execute.

Here's the set of assembly language instructions that the designer might create for the simple microprocessor in our example:

- LOADA mem - Load register A from memory address
- LOADB mem - Load register B from memory address
- CONB con - Load a constant value into register B
- SAVEB mem - Save register B to memory address
- SAVEC mem - Save register C to memory address
- ADD - Add A and B and store the result in C
- SUB - Subtract A and B and store the result in C
- MUL - Multiply A and B and store the result in C
- DIV - Divide A and B and store the result in C
- COM - Compare A and B and store the result in test
- JUMP addr - Jump to an address
- JEQ addr - Jump, if equal, to address
- JNEQ addr - Jump, if not equal, to address
- JG addr - Jump, if greater than, to address
- JGE addr - Jump, if greater than or equal, to address
- JL addr - Jump, if less than, to address
- JLE addr - Jump, if less than or equal, to address
- STOP - Stop execution

If we have read How C Programming Works, then we know that this simple piece of C code will calculate the factorial of 5 (where the factorial of 5 = 5! = 5 * 4 * 3 * 2 * 1 = 120):

```
a = 1;
f = 1;
while (a <= 5)
{
f = f * a;
a = a + 1;
}
```

At the end of the program's execution, the variable f contains the factorial of 5.

Assembly Language

A C compiler translates this C code into assembly language. Assuming that RAM starts at address 128 in this processor, and ROM (which contains the assembly language programme) starts at address 0, then for our simple microprocessor the assembly language might look like this:

```
// Assume a is at address 128
// Assume F is at address 129
0 CONB 1// a=1;
1 SAVEB 128
2 CONB 1// f=1;
3 SAVEB 129
4 LOADA 128// if a > 5 the jump to 17
5 CONB 5
6 COM
7 JG 17
8 LOADA 129// f=f*a;
9 LOADB 128
10 MUL
11 SAVEC 129
12 LOADA 128// a=a+1;
13 CONB 1
14 ADD
15 SAVEC 128
16 JUMP 4// loop back to if
17 STOP
```

ROM

So now the question is, "How do all of these instructions look in ROM?" Each of these assembly language instructions must be represented by a binary number. For the sake of simplicity, let's assume each assembly language instruction is given a unique number, like this:

- LOADA - 1
- LOADB - 2

- CONB - 3
- SAVEB - 4
- SAVEC mem - 5
- ADD - 6
- SUB - 7
- MUL - 8
- DIV - 9
- COM - 10
- JUMP addr - 11
- JEQ addr - 12
- JNEQ addr - 13
- JG addr - 14
- JGE addr - 15
- JL addr - 16
- JLE addr - 17
- STOP - 18

The numbers are known as opcodes. In ROM, our little programme would look like this:

```
// Assume a is at address 128
// Assume F is at address 129
Addr opcode/value
0 3// CONB 1
1 1
2 4// SAVEB 128
3 128
4 3// CONB 1
5 1
6 4// SAVEB 129
7 129
8 1// LOADA 128
9 128
10 3// CONB 5
11 5
12 10// COM
13 14// JG 17
14 31
15 1// LOADA 129
16 129
```

```
17 2// LOADB 128
18 128
19 8// MUL
20 5// SAVEC 129
21 129
22 1// LOADA 128
23 128
24 3// CONB 1
25 1
26 6// ADD
27 5// SAVEC 128
28 128
29 11// JUMP 4
30 8
31 18// STOP
```

We can see that seven lines of C code became 18 lines of assembly language, and that became 32 bytes in ROM.

Decoding

The instruction decoder needs to turn each of the opcodes into a set of signals that drive the different components inside the microprocessor. Let's take the ADD instruction as an example and look at what it needs to do:

- During the first clock cycle, we need to actually load the instruction. Therefore the instruction decoder needs to:
  - Activate the tri-state buffer for the programme counter
  - Activate the RD line
  - Activate the data-in tri-state buffer
  - Latch the instruction into the instruction register
- During the second clock cycle, the ADD instruction is decoded. It needs to do very little:
  - set the operation of the ALU to addition
  - latch the output of the ALU into the C register
- During the third clock cycle, the programme counter is incremented (in theory this could be overlapped into the second clock cycle).

Every instruction can be broken down as a set of sequenced operations like these that manipulate the components of the microprocessor in the proper order. Some instructions, like this ADD instruction, might take two or three clock cycles. Others might take five or six clock cycles.

MICROPROCESSOR PERFORMANCE AND TRENDS

The number of transistors available has a huge effect on the performance of a processor. As seen earlier, a typical instruction in a processor like an 8088 took 15 clock cycles to execute. Because of the design of the multiplier, it took approximately 80 cycles just to do one 16-bit multiplication on the 8088. With more transistors, much more powerful multipliers capable of single-cycle speeds become possible.

More transistors also allow for a technology called pipelining. In a pipelined architecture, instruction execution overlaps. So even though it might take five clock cycles to execute each instruction, there can be five instructions in various stages of execution simultaneously. That way it looks like one instruction completes every clock cycle.

Many modern processors have multiple instruction decoders, each with its own pipeline. This allows for multiple instruction streams, which means that more than one instruction can complete during each clock cycle. This technique can be quite complex to implement, so it takes lots of transistors.

TRENDS

The trend in processor design has primarily been toward full 32-bit ALUs with fast floating point processors built in and pipelined execution with multiple instruction streams. The newest thing in processor design is 64-bit ALUs, and people are expected to have these processors in their home PCs in the next decade.

There has also been a tendency toward special instructions (like the MMX instructions) that make certain operations particularly efficient, and the addition of hardware virtual memory support and L1 caching on the processor chip. All of these trends push up the transistor count, leading to the multi-million transistor powerhouses available today. These processors can execute about one billion instructions per second!

64-bit Microprocessors

Sixty-four-bit processors have been with us since 1992, and in the 21st century they have started to become mainstream. Both Intel and AMD have introduced 64-bit chips, and the Mac G5 sports a 64-bit processor. Sixty-four-bit processors have 64-bit ALUs, 64-bit registers, 64-bit buses and so on.

Fig. Microprocessors

One reason why the world needs 64-bit processors is because of their enlarged address spaces. Thirty-two-bit chips are often constrained to a maximum of 2 GB or 4 GB of RAM access. That sounds like a lot, given that most home computers currently use only 256 MB to 512 MB of RAM. However, a 4-GB limit can be a severe problem for server machines and machines running large databases.

And even home machines will start bumping up against the 2 GB or 4 GB limit pretty soon if current trends continue. A 64-bit chip has none of these constraints because a 64-bit RAM address space is essentially infinite for the foreseeable future — 2^64 bytes of RAM is something on the order of a billion gigabytes of RAM.

With a 64-bit address bus and wide, high-speed data buses on the motherboard, 64-bit machines also offer faster I/O (input/output) speeds to things like hard disk drives and video cards. These features can greatly increase system performance. Servers can definitely benefit from 64 bits, but what about normal users? Beyond the RAM solution, it is not clear that a 64-bit chip offers "normal users" any real, tangible benefits at the moment.

They can process data (very complex data features lots of real numbers) faster. People doing video editing and people doing photographic editing on very large images benefit from this kind of computing power. High-end games will also benefit, once they are re-coded to take advantage of 64-bit features. But the average user who is reading e-mail, browsing the Web and editing Word documents is not really using the processor in that way.

HOW QUANTUM COMPUTERS WORK

The massive amount of processing power generated by computer manufacturers has not yet been able to quench our thirst for speed and computing capacity. In 1947, American computer engineer Howard Aiken said that just six electronic digital computers would satisfy the computing needs of the United States.

Others have made similar errant predictions about the amount of computing power that would support our growing technological needs. Of course, Aiken didn't count on the large amounts of data generated by scientific research, the proliferation of personal computers or the emergence of the Internet, which have only fueled our need for more, more and more computing power.

Will we ever have the amount of computing power we need or want? If, as Moore's Law states, the number of transistors on a microprocessor continues to double every 18 months, the year 2020 or 2030 will find the circuits on a microprocessor measured on an atomic scale. And the logical next step will be to create quantum computers, which will harness the power of atoms and molecules to perform memory and processing tasks. Quantum computers have the potential to perform certain calculations significantly faster than any silicon-based computer.

Scientists have already built basic quantum computers that can perform certain calculations; but a practical quantum computer is still years away. In this article, we'll learn what a quantum computer is and just what it'll be used for in the next era of computing.

We don't have to go back too far to find the origins of quantum computing. While computers have been around for the majority of the 20th century, quantum computing was first theorized less than 30 years ago, by a physicist at the Argonne National Laboratory. Paul Benioff is credited with first applying quantum theory to computers in 1981. Benioff theorized about creating a quantum Turing machine. Most digital computers, like the one we are using to read this article, are based on the Turing Theory.

Defining the Quantum Computer

The Turing machine, developed by Alan Turing in the 1930s, is a theoretical device that consists of tape of unlimited length that is divided into little squares. Each square can either hold a symbol (1 or 0) or be left blank. A read-write device reads these symbols and blanks, which gives the machine its instructions to perform a certain programme.

Does this sound familiar? Well, in a quantum Turing machine, the difference is that the tape exists in a quantum state, as does the read-write head. This means that the symbols on the tape can be either 0 or 1 or a superposition of 0 and 1; in other words the symbols are both 0 and 1 (and all points in between) at the same time. While a normal Turing machine can only perform one calculation at a time, a quantum Turing machine can perform many calculations at once.

Today's computers, like a Turing machine, work by manipulating bits that exist in one of two states: a 0 or a 1. Quantum computers aren't limited to two states; they encode information as quantum bits, or qubits, which can exist in superposition.

Qubits represent atoms, ions, photons or electrons and their respective control devices that are working together to act as computer memory and a processor. Because a quantum computer can contain these multiple states simultaneously, it has the potential to be millions of times more powerful than today's most powerful supercomputers.

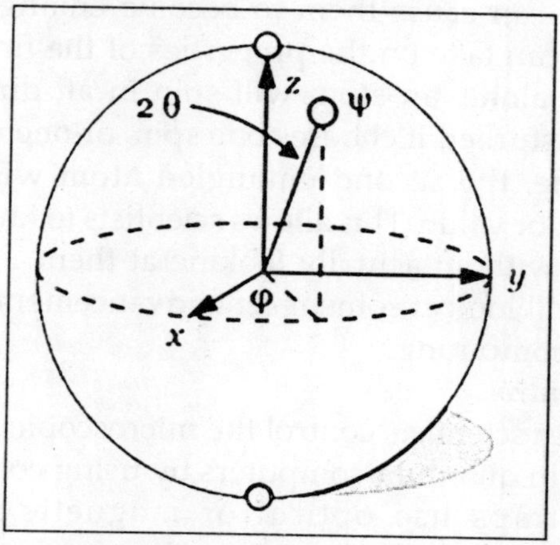

Fig. The Bloch Sphere is a Representation of a Qubit, the Fundamental Building Block of Quantum Computers

This superposition of qubits is what gives quantum computers their inherent parallelism. Parallelism allows a quantum computer to work on a million computations at once, while our desktop PC works on one. A 30-qubit quantum computer would equal the processing power of a conventional computer that could run at 10 teraflops (trillions of floating-point operations per second). Today's typical desktop computers run at speeds measured in gigaflops (billions of floating-point operations per second).

Quantum computers also utilize another aspect of quantum mechanics known as entanglement. One problem with the idea of quantum computers is that if we try to look at the subatomic particles, we could bump them, and thereby change their value.

If we look at a qubit in superposition to determine its value, the qubit will assume the value of either 0 or 1, but not both (effectively turning our spiffy quantum computer into a mundane digital computer).

To make a practical quantum computer, scientists have

to devise ways of making measurements indirectly to preserve the system's integrity. Entanglement provides a potential answer. In quantum physics, if we apply an outside force to two atoms, it can cause them to become entangled, and the second atom can take on the properties of the first atom.

So if left alone, an atom will spin in all directions. The instant it is disturbed it chooses one spin, or one value; and at the same time, the second entangled atom will choose an opposite spin, or value. This allows scientists to know the value of the qubits without actually looking at them.

Next, we'll look at some recent advancements in the field of quantum computing.

Qubit Control

Computer scientists control the microscopic particles that act as qubits in quantum computers by using control devices.

- Ion traps use optical or magnetic fields (or a combination of both) to trap ions.
- Optical traps use light waves to trap and control particles.
- Quantum dots are made of semiconductor material and are used to contain and manipulate electrons.
- Semiconductor impurities contain electrons by using "unwanted" atoms found in semiconductor material.
- Superconducting circuits allow electrons to flow with almost no resistance at very low temperatures.

Today's Quantum Computers

Quantum computers could one day replace silicon chips, just like the transistor once replaced the vacuum tube. But for now, the technology required to develop such a quantum computer is beyond our reach. Most research in quantum computing is still very theoretical.

The most advanced quantum computers have not gone beyond manipulating more than 16 qubits, meaning that they are a far cry from practical application. However, the potential remains that quantum computers one day could perform, quickly and easily, calculations that are incredibly time-consuming on conventional computers. Several key advancements have been made in quantum computing in the last few years. Let's look at a few of the quantum computers that have been developed.

1998

Los Alamos and MIT researchers managed to spread a single qubit across three nuclear spins in each molecule of a liquid solution of alanine (an amino acid used to analyse quantum state decay) or trichloroethylene (a chlorinated hydrocarbon used for quantum error correction) molecules. Spreading out the qubit made it harder to corrupt, allowing researchers to use entanglement to study interactions between states as an indirect method for analyzing the quantum information.

2000

In March, scientists at Los Alamos National Laboratory announced the development of a 7-qubit quantum computer within a single drop of liquid. The quantum computer uses nuclear magnetic resonance (NMR) to manipulate particles in the atomic nuclei of molecules of trans-crotonic acid, a simple fluid consisting of molecules made up of six hydrogen and four carbon atoms. The NMR is used to apply electromagnetic pulses, which force the particles to line up. These particles in positions parallel or counter to the magnetic field allow the quantum computer to mimic the information-encoding of bits in digital computers.

Researchers at IBM-Almaden Research Centre developed what they claimed was the most advanced quantum computer to date in August. The 5-qubit quantum computer was designed to allow the nuclei of five fluorine atoms to interact with each other as qubits, be programmed by radio frequency pulses and be detected by NMR instruments similar to those used in hospitals.

Led by Dr. Isaac Chuang, the IBM team was able to solve in one step a mathematical problem that would take conventional computers repeated cycles. The problem, called order-finding, involves finding the period of a particular function, a typical aspect of many mathematical problems involved in cryptography.

2001

Scientists from IBM and Stanford University successfully demonstrated Shor's Algorithm on a quantum computer. Shor's Algorithm is a method for finding the prime factors of numbers (which plays an intrinsic role in cryptography). They used a 7-qubit computer to find the factors of 15. The computer correctly deduced that the prime factors were 3 and 5.

2005

The Institute of Quantum Optics and Quantum Information at the University of Innsbruck announced that scientists had created the first qubyte, or series of 8 qubits, using ion traps.

2006

Scientists in Waterloo and Massachusetts devised methods for quantum control on a 12-qubit system. Quantum control becomes more complex as systems employ more qubits.

2007

Canadian startup company D-Wave demonstrated a 16-qubit quantum computer. The computer solved a sudoku puzzle and other pattern matching problems. The company claims it will produce practical systems by 2008. Skeptics believe practical quantum computers are still decades away, that the system D-Wave has created isn't scaleable, and that many of the claims on D-Wave's Web site are simply impossible (or at least impossible to know for certain given our understanding of quantum mechanics).

If functional quantum computers can be built, they will be valuable in factoring large numbers, and therefore extremely useful for decoding and encoding secret information. If one were to be built today, no information on the Internet would be safe.

Fig. D-Wave's 16-Qubit Quantum Computer

Our current methods of encryption are simple compared to the complicated methods possible in quantum computers. Quantum computers could also be used to search large databases in a fraction of the time that it would take a conventional computer. Other applications could include using quantum computers to study quantum mechanics, or even to design other quantum computers.

But quantum computing is still in its early stages of development, and many computer scientists believe the technology needed to create a practical quantum computer is years away. Quantum computers must have at least several dozen qubits to be able to solve real-world problems, and thus serve as a viable computing method.

HOW FABRIC PCS WILL WORK

Imagine walking to school or work with a brand-new type of laptop computer in hand. We walk casually, swinging the laptop back and forth between our arms, which is easy, since it weighs well under one pound (0. 45 kg) and isn't much thicker than a checkbook. Although it has no carrying case, we hardly blink after dropping it onto the concrete sidewalk. Instead we pick it up, dust it off, and continue on our way. When we arrive at our desk, we toss the laptop down on the table and open it up. The screen immediately unfolds, spreading out into an enormous display!

While this scenario sounds very futuristic, it actually isn't that far from reality, thanks in part to a concept design called a Fabric PC (personal computer), produced by Fujistu, Inc. Amazingly, a Fabric PC won't be encased within a tough metal shell like the PCs that have been around up to this point. Instead, it will be built into a tough but pliable fabric frame. Even the screen will able to bend and fold and will have a thickness and flexibility similar to a laminated sheet of paper.

Fig. Fujitsu's Fabric PC concept

Ho-w exactly will the Fabric PC concept become a reality, given its radical departure from current designs? Furthermore, how could it be predicted to hit the market in just a few years? Experts say it will be available in 2 to 10 years. Find out why this technology really is right around the corner and why some of its seemingly futuristic features aren't actually so different from what's being offered by products available on the market today.

Keep reading to lear-n how a computer could be built within a flexible fabric frame instead of a more traditional rigid metal housing.

FLEXIBLE COMPUTER

To understand how a laptop computer could be designed to be flexible, let's first consider why conventional laptops aren't flexible. To do this, we'll need to understand how current laptops are structured.

Basically, the laptops with which we're all familiar consist of many individual components, each fixed within a metal frame. Not only does a traditional laptop's metal frame form a rigid shell, but most of its internal components are bulky and rigid, too. The largest of these components is the screen, or display.

There are several different types of modern computer displays, each built with different technologies, but most of today's displays are inflexible, bulky and surprisingly fragile. The hard disk drive, used to store memory, is another large and rigid component of a traditional laptop. The list continues and includes many other components we're probably familiar with, such as the CD/DVD drive and the battery.

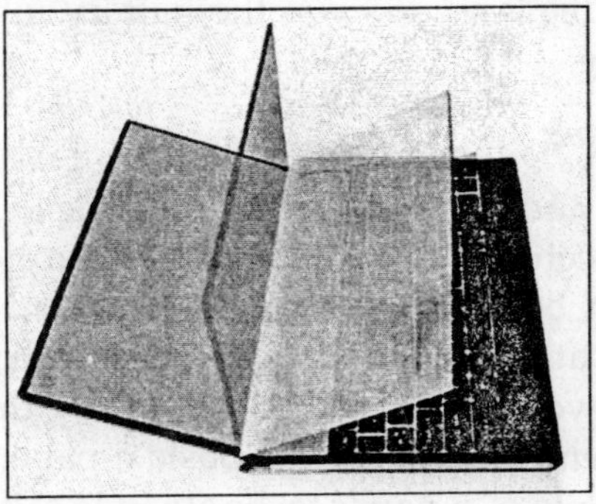

Fig. The Fabric PC is Completely Flexible — it even has a fabric Keyboard

How could a Fabric PC get around the need to be rigid, with multiple bulky components all fixed into a metal frame? There are at least three main components to the Fabric PC design that will allow this to happen. First of all, the computer display will be constructed from a cutting-edge technology called e-paper, allowing the display screen to be paper-thin and bendable.

A second essential feature of the Fabric PC's flexible design is that its individual components will be built into a pliable fabric-like backbone rather than a metal frame. While the components themselves will still be rigid, the device as a whole will have some flexibility if the components are spaced apart within the fabric — somewhat similar to how rigid rhinestones can be sewn into a fancy dress.

One final key to the Fabric PC's design is straightforward — its individual components will be as small and lightweight as possible. Also, non-essential components will likely be left out. This minimalist approach to computer design has already been the trend for some of today's laptops that are designed to be small and lightweight.

For example, lighter "flash" memory has been used in place of bulkier hard disk drives. With the ability to quickly download files wirelessly, the CD/DVD drive can be left out entirely. The e-paper display that Fabric PCs will use potentially has much lower power requirements than conventional displays, so even the battery might be a target for downsizing.

Electronic Paper

For a thin and bendable display, Fabric PCs will rely on a cutting-edge technology called e-paper, or electronic paper. The technology behind e-paper was pioneered in the 1970s by Nick Sheridan at the Xerox Palo Alto Research Centre and has continued to evolve since that time. Today there are several different implementations of the basic e-paper concept.

One example of e-paper technology is called E ink, made by the E ink Corporation. We'll very briefly touch here on how E Ink technology works, but to learn much more about it, read How Electronic Ink Works. Basically, this form of e-paper is created by sandwiching millions of tiny plastic wells between two sheets of flexible plastic. Each well contains both white and black particles, suspended within a clear fluid.

The key to this technology is that the white and black particles have opposite charges, so when an electric voltage is applied to individual wells — through circuitry embedded underneath — the black and white particles can be separated to opposite sides. In this way, the face-up side of each well can be set to appear either as black or white as seen through the top layer of clear plastic. Each well functions as a separate pixel on the E ink display.

The prototype model of an electronic paper display that is flexible and does not blur even if it is bent or pressed by a finger. e-paper, based on this type of design, can be curled or even folded like a sheet of laminated paper, and because of its light weight and flexibility it's also much less fragile than traditional displays.

Fujitsu's Fabric PC concept designs take full advantage of these characteristics by incorporating large displays that can fold up to fit within a Fabric PC's case when it is closed. The power requirements for e-paper displays are also much lower than for traditional displays. For Fabric PC's, this will translate into longer battery life and/or smaller batteries.

Keep in mind that the Fabric PC is currently just a concept design and that even working prototypes have yet to be developed. One of the factors that Fabric PC development will depend upon is continued progress in e-paper technology. As exciting as e-paper is, it's important to stress that e-paper technology is still a work in progress, too.

There are considerable technical challenges that must be overcome before low-cost e-paper displays will be available. For example, just one of those challenges will be the ability to display a full range of colors that can also update quickly enough to accommodate video output. Bringing the cost of e-paper displays down to an affordable level will be especially important as well.

FUTURE PC TECHNOLOGIES

As the technologies behind the Fabric PC concept continue to develop, durable, pliable and light-weight computer-based devices will be introduced into all corners of the business and entertainment world.

To illustrate, Fujitsu has unveiled several other concept devices along with the Fabric PC, which rely on similar basic technology.

One example is the "clerk browser" — a computer worn around the wrist like a large bracelet. This concept device was demonstrated as a means by which store clerks or salespeople could provide on-the-spot information to customers — although it's easy to imagine other applications for this type of wearable computer. As with the Fabric PC, e-paper and small, lightweight components would be essential features of this type of device.

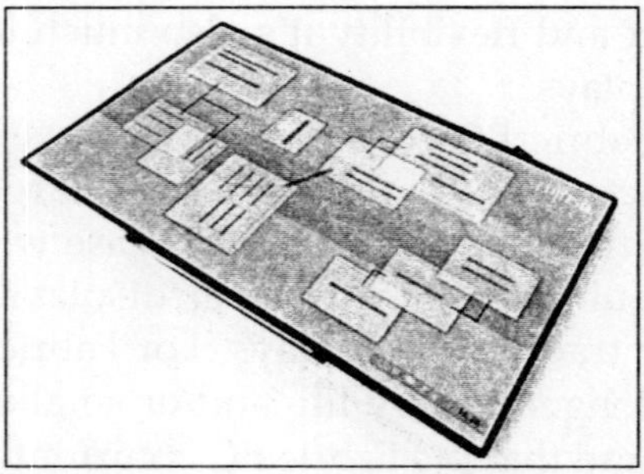

Fig. The Fabric PC has a large, flexible e-paper display.

The "card viewer" is another one of Fujitsu's concept designs that would rely on e-paper. In fact, this device can be thought of as an e-paper notepad. Because e-paper requires inputs to make changes to its display, once the display is set, it remains set, even when disconnected from its power source.

The card viewer concept takes advantage of this situation by incorporating one or more sheets of e-paper displays that can be detached from the device and handed off just like a business card. The detached sheet of e-paper will continue to display the same image — just like a normal piece of paper pulled from a notepad — until reattached and updated.

There are countless other possible applications for the types of technology that Fabric PCs will employ. Just like Fabric PCs, most of these concepts will have to wait a few years until the supporting technology becomes more advanced and affordable. In the meantime we'll have to make do with these exciting concept models, and a little imagination, too.

HOW DNA COMPUTERS WILL WORK

Even as we read this article, computer chip manufacturers are furiously racing to make the next microprocessor that will topple speed records. Sooner or later, though, this competition is bound to hit a wall. Microprocessors made of silicon will eventually reach their limits of speed and miniaturization. Chip makers need a new material to produce faster computing speeds.

We won't believe where scientists have found the new material they need to build the next generation of microprocessors.

Millions of natural supercomputers exist inside living organisms, including our body. DNA (deoxyribonucleic acid) molecules, the material our genes are made of, have the potential to perform calculations many times faster than the world's most powerful human-built computers. DNA might one day be integrated into a computer chip to create a so-called biochip that will push computers even faster. DNA molecules have already been harnessed to perform complex mathematical problems.

While still in their infancy, DNA computers will be capable of storing billions of times more data than our personal computer. In this article, we'll learn how scientists are using genetic material to create nano-computers that might take the place of silicon-based computers in the next decade.

DNA Computing Technology

DNA computers can't be found at our local electronics store yet. The technology is still in development, and didn't even exist as a concept a decade ago. In 1994, Leonard Adleman introduced the idea of using DNA to solve complex mathematical problems. Adleman, a computer scientist at the University of Southern California, came to the conclusion that DNA had computational potential after reading the book "Molecular Biology of the Gene," written by James Watson, who co-discovered the structure of DNA in 1953. In fact, DNA is very similar to a computer hard drive in how it stores permanent information about our genes.

Adleman is often called the inventor of DNA computers. The journal Science outlined how to use DNA to solve a well-known mathematical problem, called the directed Hamilton Path problem, also known as the "traveling salesman" problem. The goal of the problem is to find the shortest route between a number of cities, going through each city only once. As we add more cities to the problem, the problem becomes more difficult. Adleman chose to find the shortest route between seven cities.

We could probably draw this problem out on paper and come to a solution faster than Adleman did using his DNA test-tube computer. Here are the steps taken in the Adleman DNA computer experiment:

- Strands of DNA represent the seven cities. In genes, genetic coding is represented by the letters A, T, C and G. Some sequence of these four letters represented each city and possible flight path.
- These molecules are then mixed in a test tube, with some of these DNA strands sticking together. A chain of these strands represents a possible answer.
- Within a few seconds, all of the possible combinations of DNA strands, which represent answers, are created in the test tube.

- Adleman eliminates the wrong molecules through chemical reactions, which leaves behind only the flight paths that connect all seven cities.

The success of the Adleman DNA computer proves that DNA can be used to calculate complex mathematical problems. However, this early DNA computer is far from challenging silicon-based computers in terms of speed. The Adleman DNA computer created a group of possible answers very quickly, but it took days for Adleman to narrow down the possibilities. Another drawback of his DNA computer is that it requires human assistance. The goal of the DNA computing field is to create a device that can work independent of human involvement.

Three years after Adleman's experiment, researchers at the University of Rochester developed logic gates made of DNA. Logic gates are a vital part of how our computer carries out functions that we command it to do.

These gates convert binary code moving through the computer into a series of signals that the computer uses to perform operations. Currently, logic gates interpret input signals from silicon transistors, and convert those signals into an output signal that allows the computer to perform complex functions.

The Rochester team's DNA logic gates are the first step toward creating a computer that has a structure similar to that of an electronic PC. Instead of using electrical signals to perform logical operations, these DNA logic gates rely on DNA code. They detect fragments of genetic material as input, splice together these fragments and form a single output.

For instance, a genetic gate called the "And gate" links two DNA inputs by chemically binding them so they're locked in an end-to-end structure, similar to the way two Legos might be fastened by a third Lego between them. The researchers believe that these logic gates might be combined with DNA microchips to create a breakthrough in DNA computing.

DNA computer components — logic gates and biochips — will take years to develop into a practical, workable DNA computer. If such a computer is ever built, scientists say that it will be more compact, accurate and efficient than conventional computers.

Silicon vs. DNA Microprocessors

Silicon microprocessors have been the heart of the computing world for more than 40 years. In that time, manufacturers have crammed more and more electronic devices onto their microprocessors. In accordance with Moore's Law, the number of electronic devices put on a microprocessor has doubled every 18 months.

Moore's Law is named after Intel founder Gordon Moore, who predicted in 1965 that microprocessors would double in complexity every two years. Many have predicted that Moore's Law will soon reach its end, because of the physical speed and miniaturization limitations of silicon microprocessors.

DNA computers have the potential to take computing to new levels, picking up where Moore's Law leaves off. There are several advantages to using DNA instead of silicon:

- As long as there are cellular organisms, there will always be a supply of DNA.
- The large supply of DNA makes it a cheap resource.
- Unlike the toxic materials used to make traditional microprocessors, DNA biochips can be made cleanly.
- DNA computers are many times smaller than today's computers.

DNA's key advantage is that it will make computers smaller than any computer that has come before them, while at the same time holding more data. One pound of DNA has the capacity to store more information than all the electronic computers ever built;- and the computing power of a teardrop-sized DNA computer, using the DNA logic gates, will be more powerful than the world's most powerful supercomputer. More than 10 trillion DNA molecules can fit into an area no larger than 1 cubic centimeter (0. 06 cubic inches).

With this small amount of DNA, a computer would be able to hold 10 terabytes of data, and perform 10 trillion calculations at a time. By adding more DNA, more calculations could be performed.

Unlike conventional computers, DNA computers perform calculations parallel to other calculations. Conventional computers operate linearly, taking on tasks one at a time. It is parallel computing that allows DNA to solve complex mathematical problems in hours, whereas it might take electrical computers hundreds of years to complete them.

The first DNA computers are unlikely to feature word processing, e-mailing and solitaire programs. Instead, their powerful computing power will be used by national governments for cracking secret codes, or by airlines wanting to map more efficient routes. Studying DNA computers may also lead us to a better understanding of a more complex computer — the human brain.

HOW PRINTABLE COMPUTERS WILL WORK

Few industries can boast the huge leaps in technology that the computer industry has made in the last 50 years. Since the invention of the transistor in the 1940s, computers have shrunk from behemoth machines that took up multiple rooms to portable devices the size of paperback books that can perform hundreds of millions of operations per second. While computer manufacturing has made great strides in the last half-century, the manufacturing process is still limited to a handful of companies.

Manufacturing computers is a costly and time-consuming undertaking. A microprocessor fabrication plant costs $2 billion and takes two full weeks to produce one silicon-based microprocessor.

Few computer enthusiasts have the resources to make their own computer chips. However, researchers are developing ways to allow anyone to become their own microprocessor fabricator. Users will simply download microchip designs from the Internet and print out a working ink-based, plastic processor on a desktop fabrication machine, similar to an ink jet printer.

The next phase of computing will make the users into the creators and builders of their own computer components. We will learn how desktop fabricators will allow us to print out computer components. This article also describes a few of the projects that have already succeeded in using ink jet printers to create transistors, actuators and linear-drive motors.

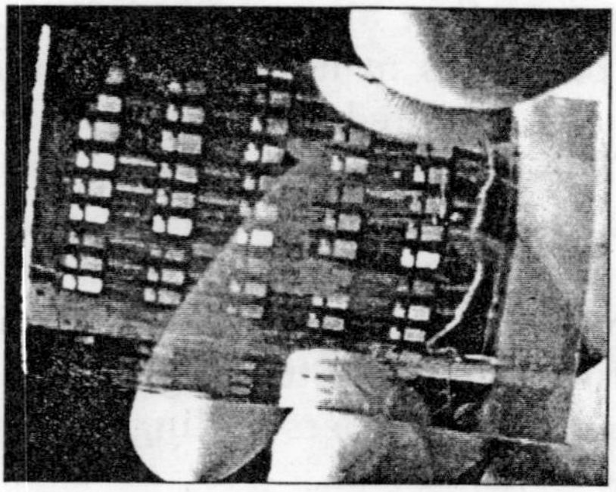

Fig. A Circuit Pattern is Printed onto Plastic using an ink-like substance

Desktop Fabrication

Few argue that the next generation of computers will be nearly invisible, meaning that they will blend in with everyday objects. Flexible ink-like circuitry will be printed onto plastic or sprayed onto various other substrates, such as clothes. One of the scientists leading this printable computer revolution is Joseph Jacobson of MIT Media Lab's Nano Media Group. Jacobson has said that his group will be able to produce a simple printed microprocessor in late 2001 or early 2002. He also foresees being able to eventually produce a printed chip that could rival an Intel Pentium processor.

Plastic Vs. Silicon

Plastic may revolutionize the semiconductor industry, but it won't be an overnight revolution. The sophistication of printable computers is still very simple. Currently, plastic fabrication devices can only produce transistors at the 25 micrometer scale (a micrometer is one-millionth of a meter); that's far from the. 2 micrometer resolution that is needed to create a working microprocessor.

Intel is able to crowd about 10 million transistors only a

few hundred nanometers big onto one silicon chip. A nanometer is one-billionth of a meter.

Most researchers will tell us that printable computer components are not designed to replace silicon. Initially, we will see these printable devices used to give intelligence to everyday objects. They will be integrated into clothes, food labels and toys. One of the most exciting applications for printable electronics is creating a wallpaper that doubles as a television screen or computer monitor. MIT also plans to build a digital camera into a business card.

Plastic does offer some benefits over silicon. Silicon is rigid, while plastic chips are flexible, allowing it to be placed on a variety of substrates. The problem is that, despite great hopes to create a plastic Pentium, printed inorganic transistors are still about 100 times slower than conventional transistors found on silicon chips.

Basically, printable computers represent the merging of conventional printing technologies with computer chip fabrication to produce cheaper, more flexible components. While many obstacles remain in its development, early products are ready to enter the market, such as disposable cell phones and computerized clothing. The next decade may bring us the ability to print out our own electronic devices and sophisticated computers.

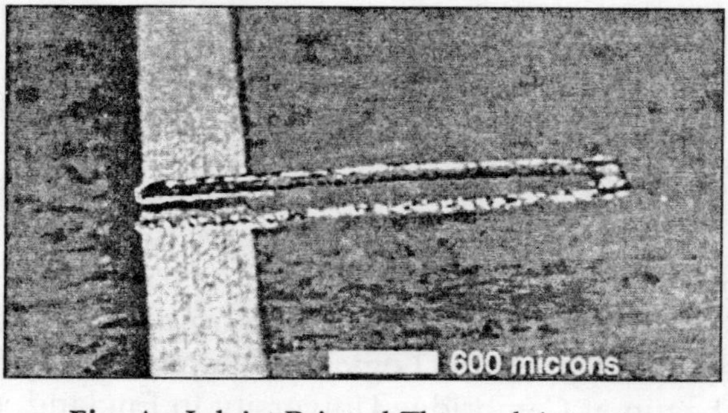

Fig. An Ink jet Printed Thermal Actuator

Jacobson's group has already succeeded in using an ordinary Hitachi ink jet printer to make several components for a printable computer. Using a nanoparticle-based ink made from suspending nano-size semiconductor particles in a liquid, researchers spray the components onto a plastic substrate. Here's a look at some of the printed components the MIT group has made with this process:

- Thermal actuators — An actuator is a sensor that causes a device to be turned on, off, adjusted or moved. In a thermal actuator, heat is used to cause the expansion of components to create movement.
- Linear-drive motors — This type of motor is similar to a normal electric motor, which has a magnet that circles around the coil loop to make the motor spin. However, there is one key difference. Think of linear-drive motors as flattened electric motors containing a flat magnet moving back and forth across a coil. In a sense, the magnet in the linear drive motor acts like a piston.
- Microelectromechanical Systems (MEMS) — MEMS are touted as the precursor or bridge to nanotechnology. These micromachines are used in a variety of devices, including pacemakers, games, and accelerometers of airbags. They perform a variety of functions, including sensing, communication and actuation. In the future, MEMS are expected to have the ability to self-replicate.

The Media Lab also created transistors using a different process. For that, polymer stamps are used with the architecture of the transistors in a positive relief. The stamp is then dipped in the nanoparticle ink and transferred to a substrate by hand. The next step will be to use an ink jet printer or some other kind of desktop fabricator to create printable transistors.

MIT isn't the only group developing ways to print computer circuitry. Plastic Logic, a company that sprang out of work begun at Cambridge University in England, plans to market the first plastic chip. The company has developed and patented a method of printing plastic onto polymer substrates, making cheap and flexible plastic transistors. The process is similar to the ink jet process used by MIT, but Plastic Logic adds carbon-based chemicals to alter the properties of the plastic. By printing the chips onto rolls of film, they can be applied to a variety of surfaces.

At Lucent Technologies' research company, Bell Labs Innovations, researchers developed the world's first printed transistor in 1997.

Using plastic sheets similar to overhead projector transparencies, a liquid plastic semiconductor is applied over a stainless-steel mesh with a squeegee to form the multiple layers of the transistor. After the solvent of the mixture evaporates, the plastic remains. The process is very similar to how silk screening works. Lucent has teamed up with E Ink, an MIT offspring, to create printable displays.

Soon, scientists will be able to create just about every part of a computer's hardware using a desktop fabricator. Plastic will take the place of silicon for many purposes, but don't expect to write off silicon as a valuable computer component for at least a decade or two.

Chapter 3

Nanotechnology Revolution

The application of nanotechnology is the practical manipulation of materials at nanoscale, placing molecules and atoms precisely in order to obtain desired functionality. Unlike working at microscale typical of most miniaturization to date, novel properties arise at nanoscale, displaying characteristics that seem inconsistent with the classical laws of physics. These novel properties can be exploited to create materials with unheralded densities and levels of integration, leading to the introduction of new uses, new devices, and new classes of applications.

Scientists and engineers need practical tools to realise the attributes and further gains possible through nanotechnology. Their tools must be able to identify and image atomic structure across a wide range of materials, and do so in a time- and -cost-effective manner. The Imago *LEAP* Microscope is the only 3D atom probe that meets all criteria.

Imago's patented Local Electrode technology led to the first practical, effective three-dimensional atom probe (3DAP), delivering statistically significant results in hours rather than days. This enables:

- A field of view 50 times larger than earlier 3DAPs; the difference between looking at the outside world through a wide-open door, instead of through a peep hole.
- Data collection a thousand times faster than conventional 3DAPs, allowing users to run dozens of experiments in a day, compared to one or two a week.

- Simplified specimen preparation and handling, with specimens made on planar surfaces such as wafers made using a variety of methods already familiar to TEM users.

What's more, Imago's state-of-the-art laser and voltage pulsers, detectors, and associated electronics deliver unparalleled precision, accuracy, and performance.

WORKING OF ATOM PROBE MICROSCOPE

Time-of-flight mass spectrometry identifies individual atoms by their elements or isotopes, while point-projection microscopy identifies where atoms were originally located in the specimen (i. e., in 3D). The atom probe microscope uses the principles of both time-of-flight mass spectroscopy and point-projection microscopy to identify individual elements and to locate them within the bulk of a material.

As in transmission electron microscopy (TEM), the user creates a specimen. Instead of the thin foil familiar to users of TEMs, the atom probe specimen is a small pointed tip (microtip) with a 100nm radius of curvature. This microtip can be carved into a wafer using commonly available techniques such as the focused ion beam. Several other techniques can be used to make the tips either before or after the film is deposited, and with or without the use of a focused ion-beam tool.

The specimen is then inserted into a cryogenically cooled, UHV analysis chamber. The analysis chamber is cryogenically cooled to freeze out atomic motion. It is at ultrahigh vacuum to allow individual atoms to be identified without interference from the environment.

A positive voltage is applied to the specimen via the voltage pulser. The positive voltage attracts electrons and results in the creation of positive ions. These ions are repelled from the specimen and pulled toward a position-sensitive detector. Under an identical voltage differential, lighter ions will reach the detector quicker than their heavier counterparts. To accurately measure the time of departure of the ion, the positive voltage is typically pulsed.

The arrival time of the ion is recorded by a set of timing devices on the single particle detector. This detector operates under the principle of a delay line detector. The time-of-flight of the ion identifies the mass-to-charge of the ion, which provides the elemental identity, so the atomic species is obtained. The numbers of atoms of various elements that hit the detector are recorded and a mass spectrum is extrapolated.

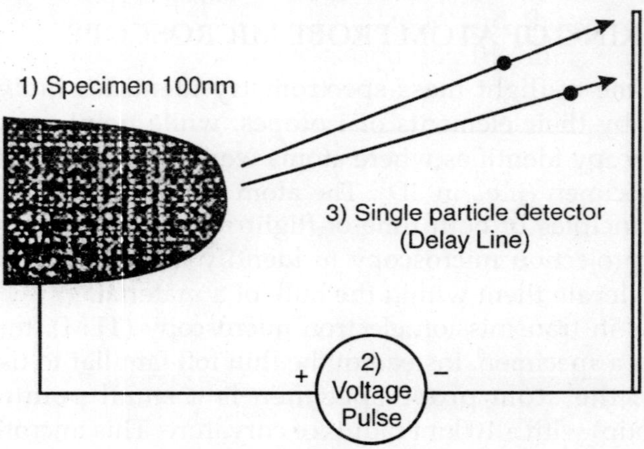

Fig. Identical Voltage Differential

The location of the atom in the specimen is determined from the ion's hit position on the delay line detector. The x and y coordinates of the hit relate to the ion's original position on the specimen. It is a fundamental rule of physics that neither potential lines nor atomic flight paths can intersect or cross each other; therefore, an ion that hits the detector to the right of another ion must have been located to its right on the specimen.

The depth, or z dimension, is provided by the sequence (time) of the ion hit on the detector. Placing the detector at a distance away from the specimen magnifies the tip of the specimen. The specimen has a curvature (r) on the order of nanometers. The detector (d) is millimeters away from the specimen. This configuration magnifies the specimen by a million X (r/d multiplied by a constant).

In due course, atoms from the surface ionize, exposing

another layer of atoms under them. This process of field ionization continues until the specimen has been fully analyzed, and provides a 3D image of the entire specimen. In this technique, the specimen is the primary optic, so there are fairly stringent considerations for the conductivity, interface strength, and geometry of the specimen.

NANO REVOLUTION

The biggest changes to shake up architecture in a long time may have their origins in the very, very small. Nanotechnology, the understanding and control of matter at a scale of one- to onehundred-billionths of a meter, is bringing incredible changes to the materials and processes of building. How ready we are to embrace them could make a big difference in the future of architectural practice.

Fig. Science of the Small

Already, this new science of the small has brought to market self-cleaning windows, smog-eating concrete, and toxin-sniffing nanosensors. Three hundred nanoengineered products are now commercially available; $32 billion worth of them were sold last year, with sales expected to top $1 trillion by 2015. But these off-the-shelf advances offer only a taste of what's incubating in the world's nanotech labs today. There, work is under way on nanocomposites thin as glass, yet capable of supporting entire buildings, and photosynthetic coatings that can make any building surface a source of free energy.

Nanotechnology works by tweaking matter from the bottom up. A nanometer is one-billionth of a meter—the paper we're reading this article on is about 100,000 nanometers thick. Recent advances in scanning electron microscopes and other technologies now make it possible to see and manipulate matter at the molecular scale more economically than ever before.

Using these tools, nanoscientists are creating revolutionary materials like coatings a single atom thick, carbon nanotubes up to 50 times stronger than steel (yet 10 times lighter), and quantum dots that could enable us to change the colour of almost any object instantaneously.

These remarkable effects are achievable because matter behaves differently at the nanoscale, where the laws of quantum physics take hold. In this quantum world, objects can change colour, shape, and phase much more easily than at the macro scale. Fundamental properties like strength, surface-to-mass ratio, conductivity, and elasticity can be engineered to create dramatically different materials.

TOMORROW'S MATERIALS

Nanotech's "wonder materials" have the potential to revolutionize how and what we build. One day, carbon nanotubes and other nanomaterials could so radically transform our material palette that paper-thin sheets might hold up entire buildings, forcing us to completely rethink the relationship between structure and skin.

Carbon nanotubes—sheets of graphite just one atom thick, formed into a cylinder—are not only 50 times stronger than steel and 10 times lighter, they are transparent and electrically conductive to boot. Nanotubes are already the building blocks for hundreds of applications, used to reinforce concrete and deliver medication to individual cells.

Nanocomposites, which combine new nanomaterials with more traditional ones such as steel, concrete, glass, and plastics, can be many times stronger than standard materials. Already on the market is a nanocomposite steel that is three times stronger than conventional steel.

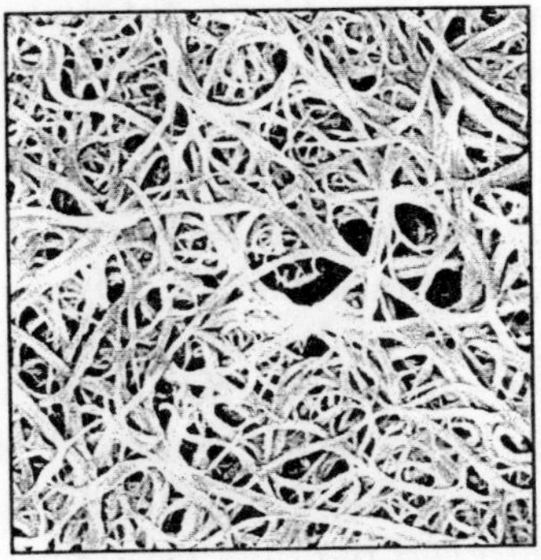

Fig. Carbon Nanotubes are up to 50 times Stronger yet 10 times Lighter than Steel

In the near term, nanocomposite reinforcement of steel, concrete, glass, and plastics will dramatically improve the performance, durability, and strength-to-weight ratio of these materials. Before long, nano-reinforced glass might be used for both structure and enclosure.

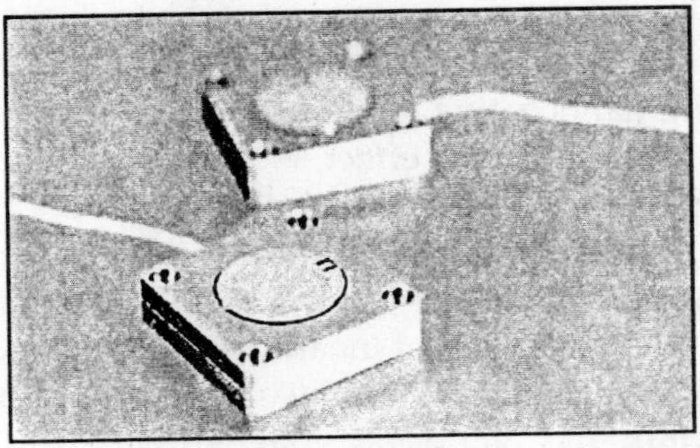

Fig. Nanosensors will Create Smart Environments—Buildings that Sense and Respond to Changes within them and to their users' Needs

In the student projects, nanotube structural panels create transparent load-bearing curtain walls free of columns and beams, quantum dots make walls and ceilings light up or change colour with the flip of a switch, and nanosensors in building components create smart environments that constantly adapt to their environment and users. But these are not just "house of the future" fantasies: My students also address the social and environmental concerns raised by nanotechnology, from toxicity (nanoparticles are so tiny, they can pass through cell membranes) to privacy (who controls the data gathered by embedded nanosensors?).

Privacy, sustainability, and security are just a few of the issues that will be profoundly affected by nanotechnology. As threats from terrorism and even from natural forces like hurricanes rise, we will utilize the strength of nanotubes to make our buildings more secure. Research that is now under way to make Army vehicle windshields bomb-proof, using polycarbonate-reinforced nanofibers, may soon be applicable to building glass.

NANOCOATINGS

Nanomaterials stand to revolutionize insulating methods because they are structured at the molecular level to trap air between particles. They are far more efficient than traditional insulators like fiberglass and polystyrene (which work at the macro level), without the environmental harm associated with those materials. And because it traps air at the molecular level, an insulating nanocoating even a few thousandths of an inch thick can have a dramatic effect.

Nanogel insulation, made by the Cabot Corp., is a form of aerogel, the lightest-weight solid in the world. Known as "frozen smoke," nanogel is 5 per cent solid and 95 per cent

air. The high air content means that a translucent panel 3. 5 inches thick can offer a high insulating value (up to R-28). Another company, Nanoseal, makes insulating paints for buildings. Its insulating coating—applied in a layer only seven-thousandths of an inch thick—is being used on beer tanks in Mexico by Corona, resulting in a temperature differential of 36 degrees Fahrenheit.

Nanocoatings are used to insulate both new and existing materials, or to protect wood, metal, and masonry, without the hazardous off-gassing of many other coatings.

Nanoengineered ultraviolet curable protective coatings by Ecology Coatings won last year's Silver Award for Innovation in *The Wall Street Journal*'s Technology Innovation Awards competition. "Garbage, paper, and other renewables can be formed into products," says Sally Ramsey, chief chemist at Ecology Coatings, "but they have a tendency to dissolve in water. We can use a very light coating of our product to waterproof these, and they can be used as a substitute for plastic. "

Good for the environment, this technology may be good for our health, too: The Hong Kong subway system has coated its cars' interiors with titanium and silver dioxide coatings that kill most of the airborne bacteria and viruses they come into contact with. And in cleansers and interior paints already used around the world, nanoparticles fight mildew.

Nanocoatings can break down dirt as well. PPG Industries and Pilkington Glass both offer self-cleaning window glass that harnesses nanotechnology. The Jubilee Church in Rome, designed by Richard Meier & Partners Architects, features self-cleaning concrete: Photocatalytic titanium dioxide nanoparticles in the precast panels (manufactured by Italcementi) make them shed dirt. The panels trap airborne pollutants in a nanoparticle matrix on their surface, then decompose them.

Similar depolluting nanocoatings can be applied to almost any surface, making it a smog-eating machine. In the near future, road surfaces, bridges, and tunnels may be able to counteract pollution. The Swedish construction giant Skanska is now involved in a $1. 7 million Swedish-Finnish project to develop catalytic cement and concrete products coated with depolluting titanium dioxide.

SMART ENVIRONMENTS

In the future, the environment will interact with occupants in ways hardly imaginable today, creating what a 2005 United Nations report calls "an internet of things. "Tiny nanosensors embedded in building materials will soon be able to track movement and detect temperature changes, humidity, toxins, weapons—even money. Sensors will pick up on users' preferences and attributes, which will then trigger responses in the intelligent objects around them, dimming the lights, altering the temperature, or—as is already happening with "push" technology that marketers use to blitz cell phones—alerting them to nearby sales and events.

Soon, the design and construction of buildings will incorporate a rich network of interacting, intelligent objects, from light-sensitive, photochromic windows to user-aware appliances. Buildings will not be static but will change constantly as their components continuously interact with users and each other. These dynamic environments will be almost organic in their ability to respond to changes, so architects will need to learn to design for change.

No longer will we call the work of design done when construction is complete. That will be only the beginning of the design process, thanks to nanotechnology.

A NANOTECHNOLOGY REVOLUTION FOR THE GEOSCIENCES

Nanotechnology uses phenomena and structures that occur on the scale of small atoms and molecules — a DNA helix, for example, is 2 nanometers in diameter — to make an array of tiny tools. Early forms of nanotechnology already pervade the modern world, in everything from paint pigments to biomedical devices.

Fig. Future Strategies for Solar Energy

Future strategies for solar energy should follow on principles in nature, where energy is stored in chemical bonds. Some promising research into this artificial photosynthesis focuses on nanotechnology for semiconductors — essentially altering solar energy cells on the molecular scale.

On the distant horizon is molecular nanotechnology, literally the organization of matter at molecular scales. Imagine, for example, "smart" clothing, in which the molecules comprising the fabric can change in response to weather. The idea, as sketched by the late Richard Feynman at a meeting of the American Physical Society at Caltech in 1959 and elaborated on later by many others, is to enable the manipulation of individual atoms and molecules, using proportionally smaller tools to build and operate even smaller tools.

Biological systems are the perfect models for molecular nanotechnology, inspiring future nanotechnology. The capabilities of biological systems put present technology to shame.

Although nanotechnology is often relegated to the field of materials science, by taking a closer look at biological systems, it could one day have a profound effect on the earth sciences. In particular, millennia-old notions of what a "resource" is, and the collection and use of energy, are both likely to change beyond recognition.

FUEL LAWS

Current technology squanders energy because most of it is used as heat. Indeed, we could speak of the "heat" crisis rather than the "energy" crisis. Fuels, after all, are burned! Two-thirds of gas in an automobile's tank, for example, goes right out the radiator. Due to the Carnot limit, a law that stems directly from fundamental constraints imposed by thermodynamics, even the most efficient heat engines waste at least half the applied energy.

Because electric batteries and motors are not heat engines, they are not subject to the Carnot limit, making them much more efficient. Conventional batteries, however, have other engineering issues such as low energy density and slow recharge times. Instead, fuel cells are a promising alternative. While similar to a battery, fuel cells allow for continuous replenishment of the reactants consumed — producing electricity from an external supply of fuel and oxygen as opposed to relying on the limited energy storage capacity of a battery. And contrary to popular belief, fuel cells do not necessarily require hydrogen.

Practical fuel cells using, say, hydrocarbons or alcohols lie beyond present technological capabilities, as converting the chemical energy of fuels directly into electricity requires a highly controlled molecular-scale reaction. This process requires catalysts that are both extremely specific and robust, and hence well structured at the nanoscale. Better catalysts in general are an obvious application of near-term nanotechnology, and will have further profound and ramifying effects on energy efficiency.

Another near-term nanotechnology solution involves solar energy. It is often claimed that the high energy density of conventional fuels is not reproducible by any conceivable alternatives, at least at the scale required for modern civilization. But the high energy density of conventional fuels is merely a brute-force solution that is compensating for the inefficiency of burning them. Thus, it is simply not true that solar power is incapable of powering a technological culture.

A high-tech culture is the only sort that can be run on solar power. After all, life itself, with its extraordinary capabilities of self-organization, synthesis and element separation, runs on solar power. That's why it is amusing to consider, for example, the oft-proposed use of biomass for fuel: Burning material originally assembled, atom by atom, from diffuse sources of both energy and materials.

ARTIFICIAL PHOTOSYNTHESIS

Why is solar power usually thought to involve converting sunlight into electricity? Biology doesn't do it that way. Natural photosynthesis stores the energy of sunlight in chemical bonds. That makes a lot more sense biologically, as well as technologically.

The conventional disadvantages of solar power are that it is intermittent, and difficult to transport and store. The last two disadvantages are true of electricity, no matter how it is created. Using sunlight to make fuels, however, would solve the intermittency problem: Fuel can accumulate whenever the sun is shining and then be used later when needed.

Fig. Acidic Drainage From mines

Acidic drainage from mines is a pollution problem that could one day be a potential resource, by using nanotechnology to separate valuable minerals and dispose of contaminants.

Artificial photosynthesis is now receiving much attention by industry and research groups. The most promising approaches are based on semiconductors — materials for controlling conductivity that make possible most of modern-day electronics, including computer chips and lasers.

As in a conventional photovoltaic (solar) cell, semiconducting materials, such as silicon, absorb solar radiation. That radiation knocks electrons loose to create a flow of current, and each excited electron leaves behind a vacancy, or "hole," that acts like a single positive charge. Conventionally, the electrons and holes are forced to drive an electric circuit before they recombine. Instead of driving an external circuit, however, technical alterations can make the hole and electron drive chemical reactions that store energy.

Most research has focused on "water splitting," the production of hydrogen gas from water, but alternative fuel generation is possible.

Familiar semiconductors, such as silicon, are too vulnerable to oxidation reactions to be used in such photochemical applications. Work instead has focused on oxide semiconductors, such as titanium dioxide, which remain stable indefinitely in an oxygen-water environment. Reaction takes place at the wetted interface between the water and semiconductor.

To be practical, however, such semiconductor surfaces will require near-molecular-scale structuring. Because the electrons and holes can combine to form heat, nanostructuring of the surface is necessary to ensure reasonable charge separation. At present, "decorating" the surface with nanoparticles of a precious metal, such as platinum, is the favored method of ensuring charge separation, but obviously this increases both expense and complexity.

An ironic result of such technologies is that desert areas, with their year-round sunlight, could become major fuel production centers. In particular, the nations of the Middle East could continue exporting fuel indefinitely, albeit in competition with other deserts throughout the world.

POLLUTANT vs RESOURCE

A fundamental technical problem involves separating one kind of atom or molecule from a background of others: pollutants from wastewater, metals from ores, salt from seawater. Separation is basic to purification, pollution control and resource extraction. Defining the process is a question of context: If we want what we separated, it is a resource; otherwise it is a pollutant.

Traditionally, however, separation has been viewed as the source of a host of different problems. In particular, researchers have seen resource extraction not only as distinct from pollution control, but also as intrinsically energy-intensive. In turn, its profligate energy usage is typically justified by vague appeals to the laws of thermodynamics.

Yet, quantitatively, element separation is not intrinsically expensive. Do not merely believe thermodynamic calculations: Bio-logical systems underscore how woefully inefficient conventional separation processes are, as they perform feats that put conventional resource extraction to shame.

Organisms do not carry out thermally driven phase separation. Instead, they literally move individual atoms or molecules, using specialized mechanisms — for example the binding of nutrient elements by specialized proteins. These molecular-scale processes are vastly less costly energetically and allow separation from considerably lower concentrations.

Plant roots extract both nutrients and water at low concentrations from the ambient soil. Vertebrate kidneys extract only certain solutes out of the blood from a background of many other solutes. For photosynthesis, plants extract carbon dioxide from the air, where its concentration is only about 350 parts per million, and furthermore do so using only the diffuse and intermittent energy of sunlight. Diatoms are particularly impressive, building shells from silica extracted at parts-per-million levels from the ambient water.

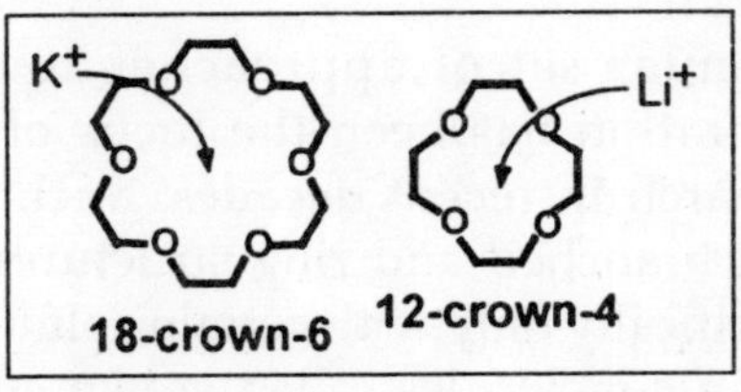

Fig. Crown Ethers

Organic compounds called "crown ethers," shown here schematically, could be key players in nanotechnology designed to extract metals. The ring, or "crown," changes in structure by substituting differently sized atoms, such as potassium or lithium, for oxygen in the crown.

Again, the reason why conventional resource extraction is so energy expensive is because it largely relies on vast quantities of heat, in this case to drive the partitioning of elements into coexisting phases. Not only do such processes require a lot of energy, but they also are intrinsically polluting, both due to the combustion necessary to generate the heat and because the separation is never complete. Moreover, byproducts containing geochemically abundant elements, such as iron in copper ores, are usually uneconomic and discarded as waste.

Thermal-based separation is also impractical for pollution control and purification. Of course, that's why such problems are traditionally viewed as distinct from resource extraction. Indeed, a number of embryonic molecular separation technologies already exist whose development has largely been driven by addressing purification issues.

EXTRACTING SOLUTIONS

In their simplest form, molecular separation techniques require that the material being separated be free to flow as a gas or a liquid. Selectivity of the separation is also fundamental: Usually only one particular dissolved species is of interest, but it is dispersed in a background of many others. Sometimes the species is valuable (for example, palladium and lithium), whereas in other cases it is toxic (for example, lead and cadmium).

One particular set of approaches toward selective molecular separation has been the focus of a tremendous amount of research in recent decades. Such efforts involve molecules with branched and ring structures that can bind tightly and specifically only with certain solutes. For instance, a group of organic molecules called crown ethers are highly effective extraction agents for many metal ions.

Crown ethers are strongly selective. The ring, or "crown," changes in structure by substituting differently sized atoms, such as nitrogen and sulfur, for oxygen in the crown. For example, the crown ether 18-crown-6 forms a strong complex with the potassium ion, which fits nicely into the ring, whereas the smaller ring of 12-crown-4 strongly binds with the lithium ion, but is too small to accommodate potassium.

One application of such a separation system is to tether

the extraction agent to a substrate to form a highly selective surface for extracting particular solutes from solution. For example, researchers have used a substituted crown ether tethered to a silica surface to recover palladium from scrap catalytic converters dissolved in acid. The palladium is bound, while the other much more abundant metals remain in solution.

The major problem with such approaches to separation is that eventually the solute must release its captured ions to regenerate the extraction agent. Typically this takes extreme chemical measures. In the palladium recovery system, for example, highly concentrated acid must be used to flush out the palladium.

Such steps generate a much larger volume of wastewater that now becomes a serious disposal problem. Separation requiring washing with fluids can be practical for recovery of highly valuable commodities like palladium, but its applications are obviously limited.

So-called switchable binding provides a way to solve the problem: Under one set of circumstances, binding occurs, but changing some variable causes the solute to unbind again. Again, biology has anticipated technology. Hemoglobin, for example, binds strongly to oxygen in the lungs, but under different chemical conditions elsewhere in the body, it gives up the oxygen to the tissues.

An example of switchable binding is "electrosorption," which is based on straightforward principles of attraction and repulsion.

Charging an electrode attracts ions with the opposite charge; reversing the charge of the electrode desorbs the ions again. Although first proposed in the 1960s for desalination, electrosorption remained impractical until the recent advent of nanostructured electrodes with very high surface areas. Because the "filled up" electrodes look like a charged capacitor, too, a great deal of the electrical energy can be recovered when the ions are desorbed.

More selective approaches require more molecular-scale structuring. For example, researchers have patented a process for extracting lithium ion from brine that uses electrodes made of a form of crystalline manganese dioxide. In the process, the electrode becomes negative, which leaves the crystal with an overall negative charge, so positive lithium cations are drawn into tunnels in the structure to compensate. Lithium cations can fit into these tunnels, whereas larger cations cannot. Reversing the charge on the manganese dioxide electrode then expels the lithium.

A similar system for extracting cesium ion is based on cesium nickel hexacyanoferrate. Here, the crystal structure contains large cavities that can accommodate the big cesium ion. Again, on applying a negative voltage, cesium cations are drawn in to compensate. The system is of great interest for extracting highly radioactive cesium-137 from nuclear waste.

An alternative potential trigger for switching binding is light. One way is to use molecules that change their structure upon absorbing a photon. The "backbone" of spiropyrans, a specialized class of molecules, for example, rearranges so drastically that a solution containing the molecule actually changes colour when illuminated. Strategically arranging the extracting groups on the backbone can make the molecule go from binding solutes in its ground state to releasing them upon illumination.

A different approach uses the absorption of light by a semiconductor surface, but such systems are nascent. In this case, the photogenerated electric charge would drive molecular mechanisms at the surface.

For example, a surface might adsorb ions from a solution in the dark, but desorb those ions when illuminated. Merely shining sunlight on a surface to desorb its solutes would obviously be much cleaner and "greener" than flushing it with strong acid solutions!

BLURRING THE LINES

At present, pollution control and purification are the key economic drivers for these separation technologies. As they mature, however, they will blur the distinction between a "pollutant" and a "resource. "Moreover, recovered pollutants will begin to have an impact on resource extraction. After all, copper extracted from a wastewater stream is copper that does not have to be mined.

Ultimately, a great many aqueous solutions, of both natural and artificial origin, will become nontraditional resources.

Wastewater streams, acid-mine drainage, seawater, concentrated natural brines such as those in oilfields or saline lakes — sometimes viewed now as problems — all could become potential sources of materials with the help of nanotechnology.

Chapter 4

Nanorobots

Nanorobotics is the technology of creating machines or robots at or close to the microscopic scale of a nanometre (10^{-9} metres). More specifically, nanorobotics refers to the still largely hypothetical nanotechnology engineering discipline of designing and building nanorobots. Nanorobots (nanobots, nanoids, nanites or nanonites) would be typically devices ranging in size from 0. 1-10 micrometers and constructed of nanoscale or molecular components. As no artificial non-biological nanorobots have yet been created, they remain a hypothetical concept.

Another definition sometimes used is a robot which allows precision interactions with nanoscale objects, or can manipulate with nanoscale resolution. Following this definition even a large apparatus such as an atomic force microscope can be considered a nanorobotic instrument when configured to perform nanomanipulation. Also macroscale robots or microrobots which can move with nanoscale precision can also be considered nanorobots.

Nanomachines are largely in the research-and-development phase, but some primitive molecular machines have been tested. An example is a sensor having a switch approximately 1. 5 nanometers across, capable of counting specific molecules in a chemical sample. The first useful applications of nanomachines, if such are ever built, might be in medical technology, where they might be used to identify cancer cells and destroy them.

Another potential application is the detection of toxic chemicals, and the measurement of their concentrations, in the environment. Recently, Rice University has demonstrated a single-molecule car which is developed by a chemical process and includes buckyballs for wheels. It is actuated by controlling the environmental temperature and by positioning a scanning tunneling microscope tip.

Nanorobots are theoretical microscopic devices measured on the scale of nanometers (1nm equals one millionth of 1 millimeter). When fully realized from the hypothetical stage, they would work at the atomic, molecular and cellular level to perform tasks in both the medical and industrial fields that have heretofore been the stuff of science fiction.

A few generations from now someone diagnosed with cancer might be offered a new alternative to chemotherapy, the traditional treatment of radiation that kills not just cancer cells but healthy human cells as well, causing hair loss, fatigue, nausea, depression, and a host of other symptoms.

A doctor practicing nanomedicine would offer the patient an injection of a special type of nanorobot that would seek out cancer cells and destroy them, dispelling the disease at the source, leaving healthy cells untouched. The extent of the hardship to the patient would essentially be a prick to the arm. A person undergoing a nanorobotic treatment could expect to have no awareness of the molecular devices working inside them, other than rapid betterment of their health.

Nanomedicine's nanorobots are so tiny that they can easily traverse the human body. Scientists report the exterior of a nanorobot will likely be constructed of carbon atoms in a diamondoid structure because of its inert properties and strength.

Super-smooth surfaces will lessen the likelihood of triggering the body's immune system, allowing the nanorobots to go about their business unimpeded. Glucose or natural body sugars and oxygen might be a source for propulsion, and the nanorobot will have other biochemical or molecular parts depending on its task.

According to current theories, nanorobots will possess at least rudimentary two-way communication; will respond to acoustic signals; and will be able to receive power or even reprogramming instructions from an external source via sound waves. A network of special stationary nanorobots might be strategically positioned throughout the body, logging each active nanorobot as it passes, then reporting those results, allowing an interface to keep track of all of the devices in the body. A doctor could not only monitor a patient's progress but change the instructions of the nanorobots *in vivo* to progress to another stage of healing. When the task is completed, the nanorobots would be flushed from the body.

Molecular nanotechnology (MNT), the umbrella science of nanomedicine, envisions nanorobots manufactured in *nanofactories* no larger than the average desktop printer. The nanofactories would use nano-scale tools capable of constructing nanorobots to exacting specifications.

Design, shape, size and type of atoms, molecules, and computerized components included would be task-specific. Raw material for making the nanorobots would be nearly cost-free, and the process virtually pollution-free, making nanorobots an extremely affordable and highly attractive technology.

The first generation of nanorobots will likely fulfill very simple tasks, becoming more sophisticated as the science progresses. They will be controlled not only through limited design functionality but also through programming and the aforementioned acoustic signaling, which can be used, notably, to turn the nanorobots off.

Robert A. Freitas Jr., author of *Nanomedicine,* gives us an example of one type of medical nanorobot he has designed that would act as a red blood cell. It consists of carbon atoms in a diamond pattern to create what is basically a tiny, spherical pressurized tank, with "molecular sorting rotors" covering just over one-third of the surface.

To make a rough analogy, these molecules would act like the paddles on a riverboat grabbing oxygen (O_2) and carbon dioxide (CO_2) molecules, which they would then pass into the inner structure of the nanorobot. The entire nanorobot which Freitas dubbed a *respirocyte,* consists of 18-billion atoms and can hold up to 9-billion O_2 and CO_2 molecules, or just over

235 times the capacity of a human red blood cell. This increased capacity is made possible because of the diamond structure supports greater pressures than a human cell.

Sensors on the nanorobot would trigger the molecular rotors to either release gasses, or collect them, depending on the needs of the surrounding tissues. A healthy dose of these nanorobots injected into a patient in solution, Freitas explains, would allow someone to comfortably sit underwater near the drain of the backyard pool for nearly four hours, or run at full speed for 15 minutes before taking a breath.

While potential medical and even military applications seem obvious for this one simple type of nanorobot, implications for every-day life are also intriguing. Imagine scuba diving without tank or regulator, but a swarm of *respirocytes* in our bloodstream; or the 2030 Olympics when, perhaps, super-athletes will not be scanned for drugs, but for nanorobotic augmentation.

Although nanorobots applied to medicine hold a wealth of promise from eradicating disease to reversing the aging process (wrinkles, loss of bone mass and age-related conditions are all treatable at the cellular level), nanorobots are also candidates for industrial applications. In great swarms they might clean the air of carbon dioxide, repair the hole in the ozone, scrub the water of pollutants, and restore our ecosystems.

PRACTICAL NANOROBOT

A nano scale medical robot is within the range of current engineering technology. It is intended for the treatment and/ or elimination of medical problems where accumulation of undesired organic substances interferes with normal bodily function, such as:

- Tumors
- Arteriosclerosis
- Blood clots leading to stroke
- Accumulation of scar tissue
- Localized pockets of infection
- Others

While much speculation has been published on possible far-future applications of nanotechnology using advanced materials and manufacturing techniques, relatively little has been published on applying existing engineering technology to the problems in order to create a solution that can be incrementally improved as the technology becomes available.

The construction and use of such devices would result in a number of benefits. Not only would it provide either cures or at least a means of controlling or reducing the effects of a number of ailments, but it will also provide valuable empirical data for the improvement and further development of such machines.

Practical data garnered from such operations at the microscopic level will allow the elimination of a number of false trails and point the way to more effective methods of dealing with the problems inherent in operation at that level.

We will address and propose solutions to problems such as size, method of entry into the body, means of propulsion, means of maintaining a fixed position while operating, control of the device, power source, means of locating substances to be eliminated,

Mans of doing the elimination and how to remove the device from the body afterward. During the course of this we will also discuss the appropriate manufacturing techniques for the construction of the device.

The problems involved in designing and building a micro-scale robot that can be introduced into the body to perform various medical activities. The preliminary design is intended for the following specific applications:

TUMORS

We must be able to treat tumors; that is to say, cells grouped in a clumped mass. While the technique may eventually be used to treat small numbers of cells in the bloodstream, this is normally done effectively by white blood cells and antibodies, and this technique is not intended to replace that.

The specified goal is to be able to destroy tumorous tissue

in such a way as to minimize the risk of causing or allowing a recurrence of the growth in the body.

The technique is intended to be able to treat tumors that cannot be accessed via conventional surgery, such as deep brain tumors. However, since the technique is extremely effective and much less debilitating than conventional surgery, it should be used, if possible, as a replacement for conventional surgery in this application.

ARTERIOSCLEROSIS

This is caused by fatty deposits on the walls of arteries. The device should be able to remove these deposits from the artery walls. This will allow for both improving the flexibility of the walls of the arteries and improving the blood flow through them. In view of the years it takes to accumulate these deposits, simply removing them from the artery walls and leaving them in the bloodstream should allow the body's natural processes to remove the overwhelming preponderance of material.

BLOOD CLOTS

The cause damage when they travel to the bloodstream to a point where they can block the flow of blood to a vital area of the body. This can result in damage to vital organs in very short order.

In many if not most cases, these blood clots are only detected when they cause a blockage and damage the organ in question, often but not always the brain. By using a microrobot in the body to break up such clots into smaller pieces before they have a chance to break free and move on their own, the chances of ensuing damage are reduced greatly.

We must consider the following factors when designing our microrobot:

- How do we introduce the device into the body?
- How do we move the device around the body?
- How do we know where the device should go?
- How do we control the device?
- How is the device powered?

- What does the device do when it gets there?
- How do we remove the device when its job is done?

DEVICE INTO THE BODY

We need to find a way of introducing the nanomachine into the body, and allowing it access to the operations site without causing too much ancillary damage. We have already made the decision to gain access via the circulatory system, which leaves us with a number of considerations.

The first is that the size of the nanomachine determines the minimum size of the blood vessel that it can traverse. Not only do we want to avoid damaging the walls of whatever blood vessel the device is in, we also do not want to block it too much, which would either cause a clot to form, or just slow or stop the blood flow, precipitating the problem we want to cure in the first place.

What this means, of course, is that the smaller the nanomachine the better. However, this must be balanced against the fact that the larger the nanomachine the more versatile and effective it can be. This is especially important in light of the fact that external control problems become much more difficult if we are trying to use multiple machines, even if they don't get in each other's way.

The second consideration is an even simpler one; we have to get it into the body without being too destructive in the first place. This requires that we gain access to a large diameter artery that can be traversed easily to gain access to most areas of the body in minimal time. The obvious candidate is the femoral artery in the leg. This is in fact the normal access point to the circulatory system for operations that require access to the bloodstream for catheters, dye injections, etc., so it will suit our purposes nicely.

DEVICE AROUND THE BODY

One of the first problems to solve is how to get our device to the problem area in the first place. We start with a basic assumption: we will use the circulatory system to allow our device to move about. We must then consider two possibilities:

should it be carried to the site of operations, or should it be propelled? We will start by dismissing the idea of using a probe, catheter or umbilicus to move the device around since this would be very difficult to make versatile enough.

The first possibility is to allow the device to be carried to the site of operations by means of normal blood flow. There are a number of requirements for this method to be practical. We must be able to navigate the bloodstream; to be able to guide the device so as to make use of the blood flow.

This also requires that there be an uninterrupted blood flow to the site of operations. In the case of tumors, there is very often damage to the circulatory system that would prevent our device from passively navigating to the site. In the case of blood clots, of course, the flow of blood is dammed and thus our device would not be carried to the site without the capability for active movement.

Another problem with this method is that it would be difficult to remain at the site without some means of maintaining position, either by means of an anchoring technique, or by actively moving against the current. While the above objections do not eliminate any possibility of ! using this technique, they do point out the need for at least a supplementary means of locomotion. There are a number of means available for active propulsion of our device.

Propeller

The very first Feynman prize in Nanotechnology was awarded to William McLellan for building an electric motor that fit within a cube 1/64th of an inch on a side. This is probably smaller than we would need for our preliminary microrobot. One or several of these motors could be used to power propellers that would push (or pull) the microrobot through the bloodstream. We would want to use a shrouded blade design so as to avoid damage to the surrounding tissues (and to the propellers) during the inevitable collisions

Cilia/Flagellae

In this scenario, we are using some sort of vibrating cilia

(similar to those of a paramecium) to propel the device. A variation of this method would be to use a fin-shaped appendage. While this may have its attractions at the molecular level of operation, an electric motor/propeller combination would be more practical at the scale we are talking about.

Electromagnetic Pump

This is a device with no moving parts that takes conductive fluid in at the front end and propels it out the back, in a manner similar to a ramjet, although with no minimum speed. It uses magnetic fields to do this. It would require high field strengths, which would be practical with high capacity conductors. At the scale we are talking about, room (or body) temperature ceramic superconductors are practical, making this a possibility.

Jet Pump

In this scenario, we use a pump (with moving parts) to propel blood plasma in one direction, imparting thrust in the opposite direction. This can either be done with mechanical pumps, or by means of steam propulsion, using jets of vaporized water/blood plasma.

Membrane Propulsion

A rapidly vibrating membrane can be used to provide thrust, as follows: Imagine a concave membrane sealing off a vacuum chamber, immersed in a fluid under pressure, that is suddenly tightened. This would have the effect of pushing some of the fluid away from the membrane, producing thrust in the direction toward the membrane.

The membrane would then be relaxed, causing the pressure of the fluid to push it concave again. This pressure would impart no momentum to the device, since it is balanced by the pressure on the other side of the device. At the macro scale, this thrust is not significant, but at the micro scale it is a practical means of propulsion.

Crawl along Surface

Rather than have the device float in the blood, or in

various fluids, the device could move along the walls of the circulatory system by means of appendages with specially designed tips, allowing for a firm grip without excessive damage to the tissue. It must be able to do this despite surges in the flow of blood caused by the beating of the heart, and do it without tearing through a blood vessel or constantly being torn free and swept away.

For any of these techniques to be practical, they must each meet certain requirements:

- The device must be able to move at a practical speed against the flow of blood.
- The device must be able to move when blood is pooling rather than flowing steadily.
- The device must be able to move in surges, so as to be able to get through the heart without being stuck, in the case of emergencies.
- The device must either be able to react to changes in blood flow rate so as to maintain position, or somehow anchor itself to the body so as to remain unmoving while operating.
- The device must be able to change direction laterally, so as to navigate the bloodstream.

From consideration of the above requirements, we can see that the most practical solution at present is one or more electric motors turning propellers. This solution is simple, well understood, and the technology has existed since 1960. The manufacturing techniques are relatively easy, as are methods for integrating it with the rest of the microrobot.

TYPES OF SENSORS

The next problem to consider is exactly how to detect the problem tissue that must be treated. We need two types of sensors. Long-range sensors will be used to allow us to navigate to the site of the unwanted tissue. We must be able to locate a tumor, blood clot or deposit of arterial plaque closely enough so that the use of short-range sensors is practical.

These would be used during actual operations, to allow

the device to distinguish between healthy and unwanted tissue. There are many different types of sensors, each suited for different purposes. Another important use for sensors is to be able to locate the position of the microrobot in the body. This is particularly true in the initial scenario, where we will only have one device in the body at a given time. Without any way of determining location from internal references, we need to be able to track the device by external means.

First we will examine the various possibilities for external sensors. These will be at least partially external to the microrobot, and their major purpose will be twofold. The first is to determine the location of the operations site; that is, the location of the clot, tumor or whatever is the unwanted tissue. The second purpose is to gain a rough idea of where the microrobot is in relation to that tissue. This information will be used to navigate close enough to the operations site that short range sensors will be useful.

Ultrasonic

This technique can be used in either the active or the passive mode. In the active mode, an ultrasonic signal is beamed into the body, and either reflected back, received on the other side of the body, or a combination of both. The received signal is processed to obtain information about the material through which it has passed. This method is, of course, greatly similar to those used in conventional ultrasound techniques, although they can be enhanced greatly over the current state of the art.

In the passive mode, an ultrasonic signal of a very specific pattern is generated by the microrobot. By means of signal processing techniques, this signal can be tracked with great accuracy through the body, giving the precise location of the microrobot at any time. The signal can either be continuous or pulsed to save power, with the pulse rate increasing or being switched to continuous if necessary for more detailed position information.

In the passive mode, the ultrasonic signal would be generated by means of a signal applied to a piezoelectric membrane, a technology that has been well developed for at least a decade. This will allow us to generate ultrasonic signals of relatively high amplitude and great complexity.

NMR/MRI

This technique involves the application of a powerful magnetic field to the body, and subsequent analysis of the way in which atoms within the body react to the field. It usually requires a prolonged period to obtain useful results, often several hours, and thus is not suited to real-time applications. While the performance can be increased greatly, the resolution is inherently low due to the difficulty of switching large magnetic fields quickly, and thus, while it may be suited in some cases to the original diagnosis, it is of only very limited use to us at present.

Radioactive Dye

This technique is basically one of illumination. A radioactive fluid is introduced into the circulatory system and its progress throughout the body is tracked by means of a fluoroscope or some other radiation-sensitive imaging system. The major advantage of this technique is that it follows the exact same path that our microrobot would take to reach the operations site.

By sufficiently increaing the resolution of the imaging system, and obtaining enough data to generate a three dimensional map of the route, it would provide valuable guidance information for the microrobot.

The active form of this technique would be to have a small amount of radioactive substance as part of the microrobot. This would allow its position to be tracked throughout the body at all times. Additionally, since the technique would not require the microrobot to use any power, or require a mechanism of any sort, it would greatly simplify the design of the microrobot.

While there are risks from radiation, the amount of radioactive substance used would not be even a fraction of the amount used in radioactive dye diagnosis. Additionally, as advances in electronic sensors continue, the amount of radiation needed for tracking would steadily be reduced. In fact, infrared sensing techniques are so advanced that we can fully shield the radioactive substance and merely track its heat directly.

X-ray

X-rays as a technique have their good points and bad points. On the plus side, they are powerful enough to be able to pass through tissue, and show density changes in that tissue. This makes them very useful for locating cracks and breaks in hard, dense tissue such as bones and teeth. On the other hand, they go through soft tissue so much more easily that an X-ray scan designed to show breaks in bone goes right through soft tissue without showing much detail.

On the other hand, a scan designed for soft tissue can't get through if there is any bone blocking the path of the x-rays. Another problem with x-rays is that it is very difficult to generate a narrow beam, and even if one could be generated, using it to scan an area in fine detail would necessitate prolonged exposure. Consequently, x-rays are useful only for gross diagnosis, for which several of the techniques listed above are far better suited.

Radio/Microwave/Heat

Again, these techniques (really all the same technique) can be used in both passive and active modes. The passive mode for the techniques depends on the various tissues in the body generating signals that can be detected and interpreted by external sensors. While the body does generate some very low frequency radio waves, the wavelength is so large that they are essentially useless for any sort of diagnostic purposes of the type we are interested in. The same is true of microwaves.

Recent developments, however, in the technology of infrared detection, offer great promise in potentially improving our ability to detect tumors by the increased heat they generate as a result of their increased metabolic state. This technology, however, is in its infancy. We do not know enough about how different cells in the body generate heat to be able to say how useful the technique would be. In the active mode, a signal is generated from outside the body and is treated the same way that ultrasonics or x-rays are; it is allowed to reflect from/pass through tissues and the result interpreted.

However, only infrared has a short enough wavelength to be able to provide the required image resolution for accurate and detailed navigation, and a great deal of image processing would be required to filter out the natural background signal from the body.

In order to use the technique to track the microrobot, a signal would need to be generated by the microrobot, detected outside the body, and interpreted to obtain position information. This is only practical for infrared or higher frequencies could be useful to obtain sufficiently accurate positional information. Recent advances in infrared sensing technology make this more attractive than might otherwise be the case.

We can come to the conclusion that there are two possible choices for our tracking system. We can either generate an ultrasonic signal and track that, or generate enough infrared or heat within the structure of our microrobot and track that. Of the two, the infrared technique is more practical, since there is far less problem of reflections and multi-path problems with infrared than with ultrasonics.

CONTROL ON DEVICE

Next, we consider the case of internal sensors. When we say internal sensors, we mean sensors that are an integral part of the microrobot and are used by it to make the final approach to the operation site and analyse the results of its operations. These sensors will be of two types. The first type will be used to do the final navigation.

When the device is within a short distance of the operation site, these sensors will be used to help it find the rest of the path, beyond what the external sensors can do. The second type of sensor will be used during the actual operation, to guide the microrobot to the tissue that should be removed and away from tissue that should not be removed.

Chemical

Chemical sensors can be used to detect trace chemicals in the bloodstream and use the relative concentrations of those chemicals to determine the path to take to reach the unwanted tissue. This would require several sensors so as to be able to establish a chemical gradient, and, for the same reason, would require a certain degree of physical separation between sensors.

While this is not a strict requirement, the alternative would be to try every path, and retrace a path when the blood chemicals diminish. While it is not difficult to create a solid state sensor for a given chemical, the difficulty increases greatly when the number of chemicals that must be analyzed increases. Consequently, we would probably need a series of microrobots, one for each chemical, or at least a set of replaceable sensor modules. An alternative to solid state sensors is for chemical analysis is described next.

Spectroscopic

This would involve taking continuous small samples of the surrounding tissue and analyzing them for the appropriate chemicals. This could be done either with a high-powered laser diode or by means of an electrical arc to vaporize small amounts of tissue.

The laser diode is more practical due to the difficulty of striking an arc in a liquid medium and also due to the side effects possible when sampling near nerve tissue. The diode could be pulsed at regular intervals, with an internal capacitor charging constantly so as to provide more power to the laser diode than the steady output of our power source.

TV Camera

This method involves us having a TV camera in the device and transmitting its picture outside the body to a remote control station, allowing the people operating the device to steer it. One disadvantage of this technique is the relatively high complexity of the sensors. On the other hand, solid state television sensors are an extremely well developed technology, and it should not be difficult to further develop it to the level needed. This could be combined with the laser diode at low power used for illumination.

UHF Sonar for Resolution, Texture

This technique would involve analyzing the return from an ultrasonic beam bounced off a nearby surface. This would give relatively low detail, and is dependent on the tissues in question having a sufficiently unique reflection. It would require either a great deal of onboard processing power or a very high communications bandwidth between the microrobot and the external systems; much more so than even the TV camera.

It can be seen that the best choice for short-range sensors is the spectroscopic technique, for the following reasons:

The equipment required is all solid state with no moving parts. While there is a certain power requirement, this can be met by using capacitors to store energy over a period of time and discharge it quickly. Another advantage of this technique is that simply by adding power to the diode beam we are destroying the unwanted tissue, thus combining the sensory and treatment requirements into the same equipment.

Samples of the blood plasma can be tested inside a closed chamber, which would give us the ability to do a chemical analysis that could detect a wide range of compounds rather than just one or two. Simply by doing chemical tracking with the sampling door closed, and cell analysis with the door open, we can combine both short range sensor requirements using one sensor.

MEANS OF TREATMENT

The treatment for each of the medical problems indicated is the same in general; we must remove the tissue or substance in question from the body. This can be done in one of several ways. We can break up the clump of substance and rely on the body's normal processes to eliminate it. Alternately, we can destroy the substance before allowing the body to eliminate the results. We can use the microrobot to physically remove the unwanted tissue. We can also use the microrobot to enhance other efforts being performed, and increase their effectiveness.

Physical Removal

This method can be effective in the treatment of arteriosclerosis. In this case, a blade, probe or edge of some sort can be used to physically separate deposits of plaque from the artery walls. The bloodstream would carry these deposits away, to be eliminated by the normal mechanisms of the body. Since it takes years before the plaque buildup reaches dangerous levels, the small amount not scavenged by the body can be regarded as not immediately significant.

In the case of blood clots, the situation is not so simple. In this case, it is possible that the action of physically attacking the clot could cause it to break away in large chunks, some of which could subsequently cause blockages in the blood flow. If we are going to do this, we need some means of preventing this from happening.

We can set up some mechanism to catch these blood clots and further break them up, or we can try to tap into the circulatory system downstream of the clot and filter out the pieces. It behooves us to work out a technique that will crush the blood clot into pieces too small to a danger.

We can simply send the device to the site of the arteriosclerosis or blood clot, scoop away sections of it, and have the device carry the tissue out of the body where it can be dissolved/destroyed. Repeated applications of this technique could remove most or all of a tumor with minimal destruction to the surrounding tissue and minimal spreading. In fact, proper use of the sampling box could be effective if it is large enough; rather than remove the substances, we can place them in the sample box, destroy them, and then rely on the body to eliminate the ash..

In the case of tumors, the problem is more serious. The act of physically shredding or even just breaking loose clumps of cells can result in the cancer metastasizing throughout the body. Since the mechanism of cancer spreading is unknown, this is a real danger. To filter the cancerous cells out of the blood immediately downstream of the tumor. Even if it is possible to distinguish cancerous cells from normal cells by filtering, this would not prevent the spread of tumor causing chemicals released by the ruptured cells.

In this case, something more drastic is indicated. Again, use of the sampling box might be useful, since it does destroy whatever is placed within it. Unfortunately, the act of removing cancerous cells to place them in the box could be dangerous. This leads us to the next alternative.

Physical Trauma

Another way of dealing with the unwanted tissues is by destroying them in situ. This would avoid damaging the cancerous cells and releasing chemicals into the bloodstream. In order to do this effectively, we need a means of destroying the cell without rupturing the cell wall until after it is safe. We shall consider a number of methods:

RESONANT MICROWAVES/ULTRASONICS

Rather than merely apply microwave/infrared or ultrasonic energy at random frequencies, the frequency of the energy could be applied at the specific frequencies needed to disrupt specific chemical bonds. This would allow us to make sure that the tumor producing chemicals created by cancerous cells would be largely destroyed, with the remaining amounts, if any, disposed of by the body's natural defenses.

Chemical

At first thought, chemical means do not seem to effective, since the device could not carry large quantities of chemicals, and making many round trips to a chemical reservoir would be difficult. However, further consideration will reveal several possibilities.

The first is to use chemical agents of extreme potency. Since we would be able to apply the chemicals directly to the tissue in question, the side effects would be much less than if we rely on the bloodstream to carry the chemicals. A useful but not essential feature would be that they have a greater effect on the undesired tissues than on normal tissue. While this is the method used in chemotherapy for cancer, the side effects can be extremely debilitating.

By delivering the chemicals in small amounts directly to the site where they are required, we could avoid most if not all of the side effects. This method could be used, although with lesser effect, to dissolve blood clots, and to dissolve deposits of arterial plaque. While it could be practical to release the chemicals at the operation site, it would not be practical, for time reasons, to inject the chemicals into each individual cell.

Another possibility is to introduce the chemicals directly into the bloodstream, but have our device affect the cancerous cells in such a way as to make them more susceptible than usual to the chemicals in question.

This would allow lower levels of chemotherapy to have the same effect as the normal method, reducing the side effects and the strain on the patient. There have been a number of experiments done with electrical stimulation of cell walls that seem to have such an effect. In fact, we could combine the two techniques by bringing chemicals to the site and then applying an electrical stimulus to the surrounding area, enhancing the effectiveness of the chemical.

Heat

The deleterious effect of heat on cells is well documented. Consequently, the use of heat to destroy cancerous tumors would seem to be a reasonable approach to take. There are a number of ways in which we can apply heat, each with advantages and disadvantages of their own. While the general technique is to apply relatively low levels of heat for prolonged periods of time, we can apply much higher levels for shorter periods of time to get the same effect. This is more practical for us, since the scale of our microrobot is small enough to make applying heat over a large area for a long time difficult.

Microwave

This is a popular method used in diathermy and other

techniques. Microwave radiation is directed at the cancerous cells, raising their temperature for a period of time, causing the death of the cells in question.

This is normally done by raising the temperature of the cells to just enough above body temperature to kill them after many minutes of exposure. In our case, this would require a means of generating a strong enough microwave signal in a package that is, frankly, pretty small.

Ultrasonic

An ultrasonic signal, which can be generated by a piezoelectric membrane or any other rapidly vibrating object, is directed at, and absorbed by, the cells being treated. This energy is converted to heat, raising the temperature of the cells and killing them. This has a number of advantages for us over the microwave technique, including small size and simplicity of the generator. This would not be very effective against either blood clots or arterial plaque, neither of which is very susceptible to prolonged low heat.

Electrical Resistance Heating

In this case, two electrodes would be placed in contact with a tumor, and a high electric current would be induced between the electrodes. This would literally cook the cancerous cells. It would not be very effective against arterial plaque or blood clots, neither of which is very conductive. It could, however, be used to enhance the effect of chemotherapy as well.

Laser

This would involve using a high-powered laser diode to burn away cancerous cells, arterial plaque and blood clots by vaporizing the unwanted materials. This is the method that would have the best chance of success against blood clots and arteriosclerosis as well as cancer cells. Of course, when we say heat in this case, we mean enough heat to vaporize tissue, not merely to warm it.

We can see that there is no one best way of treating the

unwanted tissue, since the method of treatment is different for each case. Rather than design a microrobot capable of all techniques, we will design a microrobot that can have any of several "treatment modules" installed on it, allowing the same basic design to be used.

POWER

One major requirement for our microrobot is, of course, power. We have to be able to get sufficient power to the microrobot to allow it to perform all of its required operations. There are two possible paths we can take for this. The first is to obtain the power from a source within the body, either by having a self-contained power supply, or by getting power from the bloodstream. The second possibility is to have power supplied from a source external to the body.

Source Within the Body

There are a number of possible mechanisms for this scenario. The basic idea is that the microrobot would carry its power supply within itself. It would need enough power to move to the site of the operation, perform its functions, which might be very power intensive, and then exit the body. There are three basic scenarios for on-board power supplies.

Body Heat

This method would use body heat to power the microrobot, in effect using the entire body as a power supply. The basic problem with this is that a power supply requires an energy gradient in order to function. In this case, we would need to areas of different temperature, so that we could set up a power flow between them. Since our microrobot would have to be mobile, and operate at full capacity in many different environments, this requirement would be difficult to fulfill.

Power from the Bloodstream

There are three possibilities for this scenario. In the first case, the microrobot would have electrodes mounted on its outer casing that would combine with the electrolytes in the blood to form a battery. This would result in a low voltage, but it would last until the electrodes were used up. The disadvantage of this method is that in the case of a clot or arteriosclerosis, there might not be enough blood flow to sustain the required power levels. Also, if the electrodes were ever embedded in anything that blocked their access to the blood, power would drop to zero and stay there. This means that a backup would be required.

The second way to get power from the bloodstream is by means of a fuel cell, or simply by burning blood chemicals. This is similar to a battery except that rather than obtain power from current flow between electrodes, we would obtain power by causing chemical reactions to take place at a controlled rate and obtaining power from this. This is much the same way that the body gets its own power by consuming fuel chemicals from the bloodstream. This has the same problem as the electrode method; it will stop working if access to the blood is blocked, or if the chemicals are not replenished.

Carry the Full Amount of Energy Required Directly Onboard

The third method is simply to carry the full amount of energy required directly onboard. The first case is one in which we use conventional chemical batteries. Unfortunately, the power to weight ratio of chemical batteries is extremely low, and a battery of such small size would be of limited use.

The second method is to use high-voltage capacitors to store a charge and use it gradually. As capacitor technology improves this may become practical, but at the moment the power to weight ratio is again too low. If we could manufacture body-temperature ceramic superconducting power storage coils (a distinct possibility, given the scale involved), this method becomes very attractive.

The third, and by far the most practical method, is to use an onboard nuclear power source. This would be relatively easy to shield given the amount of fuel involved, and it has other advantages as well. For one thing, the same radioactive material could be used for power and tracking, since the casing must be hotter than body temperature to produce power. This would have the effect of greatly reducing the complexity of the microrobot. For another, there would be no worries about running out of power, or insufficient power to get the job done. At the micro scale, shielding and power conversion are relatively easy, making this method extremely practical. The only major problem with this method is the social and political objections that would take place.

External to the Body

In this case, the power would be transmitted to the microrobot from outside the body. This can be done in a number of different ways, but it boils down to two possibilities. The first is to transmit the power by means of a physical connection, and the second, of course, is to transmit it without a physical connection.

Physical Connection

In the first case, we would need some sort of wire or cable to carry power between the microrobot and the outside power source. There are a number of problems with this approach. The first, of course, is that the wire needs to be able to reach inside the body to where the microrobot is.

This means that it must be thin enough to fit down every blood vessel that the microrobot can enter. If the wire is deployed from outside the body, the friction of the outer casing must be low enough to allow the wire to move freely within the blood vessels without cutting into the walls at any change of direction.

The wire must also be flexible enough to be able to withstand abrupt changes of direction without fatiguing, kinking or breaking. If the wire is deployed from the microrobot, we must have enough stored on the microrobot for it to be able to reach all the way to the operating site. We must also have a means of deploying the wire with! out tangles, and a means of retracting it back into the microrobot.

Of course, if the wire is strong enough, it would greatly

ease our movement problems, since the microrobot would then be deployed on a tether, with only navigational capabilities required, rather than long range movement. Similarly, removing it from the body would be greatly eased since it could simply retrace its path.

The next question is how the power would be transmitted. There are two possibilities: electricity and light. In the case of electricity, we must take several factors into account. The first is that the electricity needs a return path. This means that we must deploy a two-conductor cable, or use the body itself for the return path. Given the small amounts of power required, this is possible.

Another consideration to take into account is that due to the small diameter of the wire, there would inevitably be some heating of the wire, and therefore the surrounding tissue and this would have to be taken into account. The blood, of course, would act to carry away most of the heat. We could also use the wire for high-speed two-way communications, making that job much easier.

If the power is transmitted in the form of light, which is then either used directly or converted to electricity, the problems are different. There is no requirement for a return path, nor is there any significant leakage along the length of a fibre-optic cable of such a short length.

On the other hand, the problem of brittleness is much more significant at the diameters required. This is especially true if the fibre-optic cable is stored in, and deployed from, the microrobot itself. There is also a problem in that the conversion of light to electricity would require more on-board equipment. Of the two techniques, electricity is the better choice at this state of the art, and this is unlikely to change for this application.

NO PHYSICAL CONNECTION

In this scenario, we are transmitting power to the microrobot without the use of wires or any sort of physical means to transfer the power. Here we have a number of choices.

Microwave

In this case, there would be an antenna built into the physical structure of the microrobot. Microwave energy would be beamed into the body, where it would be picked up by the onboard antenna and converted into electricity. The first problem, of course, is readily apparent; most of the microwaves will be absorbed by the conductive body fluids, causing a great deal of ancillary damage. While microwave ovens work by agitating water molecules, and heating food that way, the fact is that any microwave beam directed at the body will be largely converted to heat.

Ultrasonic

This technique is similar to that of the microwaves, except that since water is such a good conductor of sound, most of the energy would not go into heating up the tissues in the path of the beam. Instead, they would tend to dissipate, and would be absorbed by the body as a whole, with much less attendant danger. A piezoelectric membrane would be used to pick up the ultrasonic waves and convert them to electricity. This membrane, of course, could be modulated at the same time to act as a communications device (two-way) and for a sensor device, as well.

Induced Magnetic

In this case, the body is surrounded by a magnetic field. This field would induce currents within a rotating closed conducting loop in the microrobot, which it would then use for power. The frequency of the resulting power is dependent on the rotational speed of the pickup loop, and so alternating the rotational frequency (mechanical FM modulation) would provide a communications path as well. By switching the current through a relatively high resistance path, we would obtain a pinpoint heat source, which could be used for treatment as well.

If we can maintain the physical connection, a wire deployed from the microrobot itself would be very useful, and solve many of the problems we would encounter. However, if no physical connection can be maintained, either ultrasonics or magnetic induction could be used, with ultrasonics appearing to be somewhat more effective.

Control System

We need to steer the microrobot to where the sensors tell us it needs to be. As always, the two choices are internal control and external. The following are considerations:

Need to Know Where to Go

This does not necessarily mean that we have a detailed map of the body that the microrobot is following. It simply means that the microrobot must be able to proceed to the location of the unwanted tissue within the specified time constraints, if any. If the microrobot is permanently introduced into the body to circulate and remove unwanted tissues as they are detected, this requirement is largely unnecessary.

Need to Know the Route

Some places are more difficult than others to reach. For example, a tumor deep within the brain can be located by various means, but it cannot be accessed by conventional surgical techniques. We may be able to locate the tumor by means of conventional techniques, but the sequence of blood vessels that we need to follow may be more difficult to determine. Again, for a maintenance routine, this requirement may be unnecessary as long as the microrobot covers the entire bloodstream, or at least all the sections that it can access, in a reasonable time.

Need to be Able to Correct if Drawn off Course

This requirement is only necessary if there is a specific track that must be followed to reach the necessary location. This can be true for two different scenarios.

The first is when a preplanned route exists and must be followed. Due to the complexity of the circulatory system, this will generally be the case only when the tumor can be accessed from the larger and more obvious blood vessels.

The second scenario is when the microrobot is using long range sensors, specifically chemical sensors, to locate the tumor. In this case, the microrobot would be functioning in a manner similar to a bloodhound. A reduction of the chemical trail used to locate the tumor would indicate a "wrong turn" in the bloodstream, and the need to backtrack to the point where the chemical traces started to diminish. In this case, we must also be sure not to be fooled by eddies in the bloodstream that cause a momentary reduction of the chemicals that are detected.

Need to be Able to Apply Treatment Effectively

Once we have reached the location of the tumor, clot or deposit of arterial plaque, we must be able to apply the appropriate treatment without making matters worse. We do not want to cause tumor producing chemicals or cells to scatter throughout the bloodstream. Similarly, we do not want clots to break up into large chunks, precipitating the very strokes we are trying to prevent; nor do we want to pierce the wall of an artery rather than simply remove plaque deposits.

Need to be Able to Reach outlet from Body

This is a problem if the microrobot has been introduced into the body in order to perform a specific task. In such a case, it will need to be removed, which means that either it must obtain egress from the circulatory system, or it must pass through an already existing port of exit. It can either proceed to a point where it can be removed easily, or it can backtrack to where is first entered the body.

In the case of semi-permanent introduction of the microrobot into the bloodstream for maintenance purposes, the problem is the exact opposite; we want to avoid the removal of the microrobot from the body unless it is done deliberately. One way of doing this is by means of chemical sensors. When the chemicals that accompany the breakdown of platelets and the formation of a blood clot are detected, the microrobot would swim "upstream"; away from the clot formation.

Of course, the microrobot is supposed to seek out clots in order to destroy them. We would have to find some way of distinguishing between clots that are caused by an opening in the circulatory system large enough for the microrobot to exit, and those that are small and growing only gradually. This is especially true since if we do not distinguish between them, the microrobot will constantly be breaking up clots around a wound and reopening it, causing a particularly ironic form of hemophilia.

Need to Compensate for the Unexpected

Certainly while the techniques are being developed, there will be many unexpected events. Even after the control techniques are perfected, there will be many occasions where it will be necessary to have external decision making introduced into the control loop; i. e. we are not going to be creating an autonomous microrobot any time soon. There are two ways we can handle this problem. Either the microrobot is autonomous for simple things, and calls for help when something unexpected happens, or it can be completely externally controlled, greatly reducing the complexity of the on-board processing power.

The requirements cannot even be met for a car navigating in a city, which is a similar but much less complex problem. However, not all of these requirements are entirely binding. Let us consider each one in sequence.

The only real thing that we need to know about where to go is that there is tissue to be treated along the route from introduction to egress of the microrobot. This can be accomplished in several ways. Introduction of the microrobot into the bloodstream at the correct point will allow it to move to the target by means of simply following the blood vessels appropriately. In a maintenance programme, the microrobot does not even need a destination, but simply goes where it will while trying to sense target tissues and act appropriately.

While it would be more effective to know the shortest or most effective route to the target tissue, this is only a constraint if there is a time constraint as well. For a maintenance routine, it does not matter where the microrobot is, although a chemical tracking system of some sort would improve performance greatly over random sampling. If we do not have a specific location as goal, we need not worry about the route, save that w! e do not want the microrobot to wind up in some sort of eddy with no escape. For that matter, we do not want it to wind up embedded in the wall of a vein, or in a sludge of arterial plaque. We do need some means of detecting and avoiding such an occurrence.

We can see that even though we have reduced some of the control requirements for our microrobot, the remaining considerations are well beyond the capabilities of modern programming techniques. If we had thousands or millions of nanorobots in the bloodstream, this would be a serious obstacle. However, with only a very few microrobots to control at once, we can actually (assuming sufficient communications bandwidth) have a person controlling the microrobot directly.

Means of Recovery from the Body

Given sufficiently accurate control of the nanomachine, or a tether, this is not a problem; we can just retrace our path upstream. However, it would be a lot easier, and recommended, to steer a path through the body that traverses major blood vessels and winds up at a point where we can just filter the nanomachine out of the bloodstream.

This will reduce the possibilities for difficulties, and also cause less wear and tear on the nanomachine. Of course, either scenario is a possibility, depending on where the actual operation site is. Another possibility is to have the nanomachine anchor itself to a blood vessel that is easily accessible from outside, and perform a small surgical operation to remove it.

OTHER APPLICATIONS

Kidney Stones

As anyone who has ever been plagued by kidney stones can attest, they are extremely painful, as well as being difficult to treat. In most cases, the pain must be endured until the stones have been passed. Attempts can be made to break up the stones by means of high intensity ultrasonics, but these attempts are difficult and not very successful.

By introducing a microrobot of the type described in this paper into the urethra in a manner similar to that of inserting a catheter, direct access to the kidney stones can be obtained, and they can be broken up directly. This can be done either by means of ultrasonics directly applied, or by the use of a laser or other means of applying intense local heat to cause the stones to break up. If these techniques do not work, direct physical force by means of a sintered tungsten carbide cutting or abrasive surface could be used.

Liver Stones

Liver stones accumulate in the bile duct, and while they are nowhere near as painful as kidney stones, they can still cause serious health problems. Microrobots can be introduced into the bile duct and used to break up the liver stones as well. By continuing on up the bile duct into the liver, they can clear away accumulated deposits of unwanted minerals and other substances as well. This of course, is true for the kidneys as well.

Gout

Gout occurs when the breakdown products of various fats cannot be removed from the bloodstream by the kidneys. These byproducts tend to crystallize at or near the joints, notably in the lower extremities, and cause excruciating pain to those who suffer from it. When a microrobot is in the bloodstream, it can locate these deposits by means of a combination of chemical sensors and external tracking, and can break up the crystals, allowing the bloodstream to carry them away. Of course, this will in no way prevent recurrence of the problem, but it will alleviate the symptoms for a time.

Burn and Wound Debriding

The microrobots can also be used to clean wounds and

burns. Their size allows them to be very useful for removing dirt and foreign particles from incised and punctured wounds, as well as from burns. They can be used to do a more complete and less traumatic job than conventional techniques.

Parasite Removal

The microrobots can also be used to attack other life forms in the body. For example, they would be well suited to deal with such parasites as heartworms (hopefully in pets rather than humans), liver flukes (definitely in humans). As the sensor technology improves, they could be used to attack various bacteria and other smaller organisms as well, although this would probably require the introduction of large numbers of the units into the body. In essence, this would be creating artificial antibodies, and while this is the logical extrapolation of the technology, it will not happen for some time.

Remove or Break Down tar, etc in Lungs

If the units are operating in maintenance autonomous mode, they could be very useful for the treatment of dirty lungs. This could be done by removing particles of tar and other pollutants from the surface of the alveoli, and placing them where the natural processes of the body can dispose of them. Alternatively, the unwanted substances could be vaporized or otherwise reduced to their component elements. This would require a microrobot capable of moving within the lungs, on alveolar surfaces as well as over the mucus layer and over the cilia within the lungs.

Chapter 5

Lithography

Lithography is a method for printing using a plate or stone with a completely smooth surface. By contrast, in intaglio printing plate is engraved, etched or stippled to make cavities to contain the printing ink, and in woodblock printing and letterpress ink is applied to the raised surfaces of letters or images. Lithography uses oil or fat and gum arabic to divide the smooth surface into hydrophobic regions which accept the ink, and hydrophilic regions which reject it and thus become the background. Invented by Bavarian author Alois Senefelder in 1796, it can be used to print text or artwork onto paper or another suitable material.

All types of high-volume text, are now printed using offset lithography, the most common form of printing production. The word "lithography" also refers to photolithography, a microfabrication technique used to make integrated circuits and microelectromechanical systems, although those techniques have more in common with etching than with lithography.

THE PRINCIPLE OF LITHOGRAPHY

Lithography uses chemical processes to create an image. For instance, the positive part of an image would be a hydrophobic chemical, while the negative image would be hydrophilic.

Thus, when the plate is introduced to a compatible printing ink and water mixture, the ink will adhere to the positive image and the water will clean the negative image. This allows a flat print plate to be used, enabling much longer and more detailed print runs than the older physical methods of printing (e. g., intaglio printing, Letterpress printing).

Lithography was invented by Alois Senefelder in Bohemia in 1796. In the early days of lithography, a smooth piece of limestone was used (hence the name "lithography"—"lithos" is the ancient Greek word for stone). After the oil-based image was put on the surface, a solution of gum arabic in water, was applied, the gum sticking only to the non-oily surface. During printing, water adhered to the gum arabic surfaces and avoided the oily parts, while the oily ink used for printing did the opposite.

Lithography on Limestone

Lithography works because of the mutual repulsion of oil and water. The image is drawn on the surface of the print plate with a fat or oil-based medium (hydrophobic), which may be in pigmented to make the drawing visible. A wide range of oil-based media is available, but the durability of the image on the stone depends on the lipid content of the material being used, and its ability to withstand water and acid. Following the drawing of the image, an aqueous solution of gum arabic, weakly acidified with nitric acid HNO_3 is applied to the stone.

The function of this solution is to create a hydrophilic layer of calcium nitrate salt, $Ca(NO_3)_2$, and gum arabic on all non-image surfaces. The gum solution penetrates into the pores of the stone, completely surrounding the original image with a hydrophilic layer that will not accept the printing ink. Using lithographic turpentine, the printer then removes any excess of the greasy drawing material, but a hydrophobic molecular film of it remains tightly bonded to the surface of the stone, rejecting the gum arabic and water, but ready to accept the oily ink.

When printing, the stone is kept wet with water. Naturally the water is attracted to the layer of gum and salt created by the acid wash. Printing ink based on drying oils such as linseed oil and varnish loaded with pigment is then rolled over the surface. The water repels the greasy ink but the hydrophobic areas left by the original drawing material accept it. When the hydrophobic image is loaded with ink, the stone and paper are run through a press which applies even pressure over the surface, transferring the ink to the paper and off the stone.

Senefelder had experimented in the early 1800s with multicolor lithography; he predicted that the process would eventually be perfected and used to reproduce paintings. Multi-colour printing was introduced through a new process developed by Godefroy Engelmann (France) in 1837 known as Chromolithography. A separate stone was used for each colour, and a print went through the press separately for each stone. The main challenge was of course to keep the images aligned (*in register*). This method lent itself to images consisting of large areas of flat colour, and led to the characteristic poster designs of this period.

The Modern Lithographic Process

The earliest regular use of lithography for text was in countries using Arabic, Turkish and similar scripts, where books, especially the Qu'ran, were sometimes printed by lithography in the nineteenth century, as the links between the characters require compromises when movable type is used which were considered inappropriate for sacred texts.

High-volume lithography is used today to produce posters, maps, books, newspapers, and packaging — just about any smooth, mass-produced item with print and graphics on it. Most books, indeed all types of high-volume text, are now printed using offset lithography.

In offset lithography, which depends on photographic processes, flexible aluminum, polyester, mylar or paper printing plates are used in place of stone tablets. Modern printing plates have a brushed or roughened texture and are covered with a photosensitive emulsion. A photographic negative of the desired image is placed in contact with the emulsion and the plate is exposed to ultraviolet light. After development, the emulsion shows a reverse of the negative image, which is thus a duplicate of the original (positive) image.

The image on the plate emulsion can also be created through direct laser imaging in a CTP (Computer-To-Plate) device called a platesetter. The positive image is the emulsion that remains after imaging. For many years, chemicals have been used to remove the non-image emulsion, but now plates are available that do not require chemical processing.

The plate is affixed to a cylinder on a printing press. Dampening rollers apply water, which covers the blank portions of the plate but is repelled by the emulsion of the image area. Ink, which is hydrophobic, is then applied by the inking rollers, which is repelled by the water and only adheres to the emulsion of the image area—such as the type and photographs on a newspaper page.

If this image were directly transferred to paper, it would create a negative image and the paper would become too wet. Instead, the plate rolls against a cylinder covered with a rubber *blanket,* which squeezes away the water,picks up the ink and transfers it to the paper with accurate pressure. The paper rolls across the blanket drum and the image is transferred to the paper. Because the image is first transferred, or *offset* to the rubber drum, this reproduction method is known as *offset lithography* or *offset printing*.

Many innovations and technical refinements have been made in printing processes and presses over the years, including the development of presses with multiple units (each containing one printing plate) that can print multi-colour images in one pass on both sides of the sheet, and presses that accommodate continuous rolls (*webs*) of paper, known as web presses.

Another innovation was the continuous dampening system first introduced by Dahlgren instead of the old method which is still used today on older presses (conventional dampening), which are rollers covered in molleton (cloth) which absorbs the water. This increased control over the water flow to the plate and allowed for better ink and water balance. Current dampening systems include a "delta effect or vario "which slows the roller in contact with the plate, thus creating a sweeping movement over the ink image to clean impurities known as "hickies".

The advent of desktop publishing made it possible for

type and images to be manipulated easily on personal computers for eventual printing on desktop or commercial presses. The development of digital imagesetters enabled print shops to produce negatives for platemaking directly from digital input, skipping the intermediate step of photographing an actual page layout. The development of the digital platesetter in the late twentieth century eliminated film negatives altogether by exposing printing plates directly from digital input, a process known as computer to plate printing.

MICROLITHOGRAPHY AND NANOLITHOGRAPHY

Microlithography and nanolithography refer specifically to lithographic patterning methods capable of structuring material on a fine scale. Typically features smaller than 10 micrometers are considered microlithographic, and features smaller than 100 nanometers are considered nanolithographic. Photolithography is one of these methods, often applied to semiconductor manufacturing of microchips. Photolithography is also commonly used in fabricating MEMS devices. Photolithography generally uses a pre-fabricated photomask or reticle as a master from which the final pattern is derived.

Although photolithographic technology is the most commercially advanced form of nanolithography, other techniques are also used. Some, for example electron beam lithography, are capable of much higher patterning resolution (sometime as small as a few nanometers). Electron beam lithography is also commercially important, primarily for its use in the manufacture of photomasks.

Electron beam lithography as it is usually practiced is a form of maskless lithography, in that no mask is required to generate the final pattern. Instead, the final pattern is created directly from a digital representation on a computer, by controlling an electron beam as it scans across a resist-coated substrate. Electron beam lithography has the disadvantage of being much slower than photolithography.

In addition to these commercially well-established techniques, a large number of promising microlithographic and nanolithographic technologies exist or are emerging, including nanoimprint lithography, interference lithography, X-ray lithography, extreme ultraviolet lithography, and scanning probe lithography. Some of these emerging techniques have been used successfully in small-scale commercial and important research applications. Surface-charge lithography, in fact PDMS can be directly patterned on polar dielectric crystals via pyroelectric effect, Diffraction lithography

Lithography as an Artistic Medium

During the first years of the nineteenth century, lithography made only a limited impact on printmaking, mainly because technical difficulties remained to be overcome. Germany was the main centre of production during this period. Godefroy Engelmann, who moved his press from Mulhouse to Paris in 1816, largely succeeded in resolving the technical problems, and in the 1820s lithography was taken up by artists such as Delacroix and Géricault.

London also became a centre, and some of Géricault's prints were in fact produced there. Goya in Bordeaux produced his last series of prints in lithography - *The Bulls of Bordeaux* of 1828. By the mid-century the initial enthusiasm had somewhat died down in both countries, although lithography continued to gain ground in commercial applications, which included the great prints of Daumier, published in newspapers. Rodolphe Bresdin and Jean-Francois Millet also continued to practice the medium in France, and Adolf Menzel in Germany.

In 1862 the publisher Cadart tried to launch a portfolio of lithographs by various artists which flopped, but included several superb prints by Manet. The revival began in the 1870s, especially in France with artists such as Odilon Redon, Henri Fantin-Latour and Degas producing much of their work in this way. The need for strictly limited editions to maintain the price had now been realized, and the medium become more accepted.

In the 1890s colour lithography became enormously

popular with French artists, Toulouse-Lautrec most notably of all, and by 1900 the medium in both colour and monotone was an accepted part of printmaking, although France and the US have used it more than other countries. George Bellows, Alphonse Mucha, Max Kahn, Pablo Picasso, Eleanor Coen, Jasper Johns, David Hockney and Robert Rauschenberg are a few of the artists who have produced most of their prints in the medium.

M. C. Escher is considered a master in lithography, and many of his prints were created using this process. More than other printmaking techniques, printmakers in lithography still largely depend on access to a good printer, and the development of the medium has been greatly influenced by when and where these have been established.

As a special form of lithography, the *Serilith* process is sometimes used. Serilith are mixed media original prints created in a process where an artist uses the lithograph and serigraph process. The separations for both processes are hand drawn by the artist. The serilith technique is used primarily to create fine art limited print editions.

ELECTRON BEAM LITHOGRAPHY

Electron beam lithography (often abbreviated as e-beam lithography) is the practice of scanning a beam of electrons in a patterned fashion across a surface covered with a film (called the resist), ("exposing" the resist) and of selectively removing either exposed or non-exposed regions of the resist ("developing"). The purpose is to create very small structures in the resist that can subsequently be transferred into another material for a number of purposes, for example for the creation of very small electronic devices.

The primary advantage of electron beam lithography is that it is one of the ways to beat the diffraction limit of light and make features in the nanometer regime. This form of maskless lithography has found wide usage in mask-making used in photolithography, low-volume production of semiconductor components, and research & development.

On the other hand, the key limitation of electron beam

lithography is throughput, i. e., the very long time it takes to expose an entire silicon wafer or glass substrate. A long exposure time leaves the user vulnerable to beam drift or instability which may occur during the exposure. Also, the turn-around time for reworking or re-design is lengthened unnecessarily if the pattern is not being changed the second time.

ELECTRON BEAM LITHOGRAPHY SYSTEMS

Electron beam lithography systems used in commercial applications are dedicated e-beam writing systems that are very expensive (>$4M USD). For research applications, it is very common to convert an electron microscope into an electron beam lithography system using a relatively low cost accessory (<USD 100k). Such converted systems have produced linewidths of ~20 nm since at least 1990, while current dedicated systems have produced linewidths on the order of 10 nm or smaller.

Electron beam lithography systems can be classified according to both beam shape and beam deflection strategy. Older systems used Gaussian-shaped beams and scanned these beams in a raster fashion. Newer systems use shaped beams, which may be deflected to various positions in the writing field (this is also known as vector scan).

Electron Sources

Lower resolution systems can use thermionic sources, which are usually formed from LaB_6. However, systems with higher resolution requirements need to use field emission sources, such as heated W/ZrO_2 for lower energy spread and enhanced brightness. Thermal field emission sources are preferred over cold emission sources, in spite of their slightly larger beam size, because the former offer better stability over typical writing times of several hours.

Lenses

Both electrostatic and magnetic lenses may be used. However, electrostatic lenses have more aberrations and so are not used for fine focusing. There is no current mechanism to make achromatic electron beam lenses, so extremely narrow dispersions of the electron beam energy are needed for finest focusing.

Stage, Stitching & Alignment

Typically, for very small beam deflections electrostatic deflection 'lenses' are used, larger beam deflections require electromagnetic scanning. Because of the inaccuracy and because of the finite number of steps in the exposure grid the writing field is of the order of 100 micrometre - 1 mm. Larger patterns require stage moves. An accurate stage is critical for stitching (tiling writing fields exactly against each other) and pattern overlay (aligning a pattern to a previously made one).

Electron Beam Write Time

The minimum time to expose a given area for a given dose is given by the following formula:

Dose * exposed area = beam current * exposure time/ step size **2 = total charge of incident electrons

For example, assuming an exposure area of 1 cm^2, a dose of 10^{-3} Coulombs/cm^2, and a beam current of 10^{-9} Amperes, the resulting minimum write time would be 10^6 seconds (about 12 days). This minimum write time does not include time for the stage to move back and forth, as well as time for the beam to be blanked (blocked from the wafer during deflection), as well as time for other possible beam corrections and adjustments in the middle of writing.

To cover the 700 cm^2 surface area of a 300 mm silicon wafer, the minimum write time would extend to $7*10^8$ seconds, about 22 years. This is a factor of about 10 million times slower than current optical lithography tools. It is clear that throughput is a serious limitation for electron beam lithography, especially when writing dense patterns over a large area.

E-beam lithography is not suitable for high-volume manufacturing because of its limited throughput. The serial nature of electron beam writing makes for very slow pattern generation compared with a parallel technique like photolithography (the current standard) in which the entire surface is patterned at once (1X optical steppers only, 4X or 5X steppers take proportionally longer). To pattern a single wafer with an electron beam lithography system for sub-100 nm resolution, it would typically take days, compared to the few minutes it would take with a photolithography system. Even an optical maskless lithography tool is much faster than an electron beam used for the same purpose.

Defects in Electron-Beam Lithography

Despite the high resolution of electron-beam lithography, the generation of defects during electron-beam lithography is often not considered by users. Defects may be classified into two categories: data-related defects, and physical defects.

Data-related defects may be classified further into two sub-categories. Blanking or deflection errors occur when the electron beam is not deflected properly when it is supposed to, while shaping errors occur in variable-shaped beam systems when the wrong shape is projected onto the sample. These errors can originate either from the electron optical control hardware or the input data that was taped out. As might be expected, larger data files are more susceptible to data-related defects.

Physical defects are more varied, and can include sample charging (either negative or positive), backscattering calculation errors, dose errors, fogging (long-range reflection of backscattered electrons), outgassing, contamination, beam drift and particles. Since the write time for electron beam lithography can easily exceed a day, "randomly occurring" defects are more likely to occur. Here again, larger data files can present more opportunities for defects.

Photomask defects largely originate during the electron beam lithography used for pattern definition.

ELECTRON ENERGY DEPOSITION IN MATTER

Electron trajectories in resist: An incident electron (purple) produces secondary electrons (blue). Sometimes, the incident electron may itself be backscattered as shown here and leave the surface of the resist (amber). The primary electrons in the incident beam lose energy upon entering a material through inelastic scattering or collisions with other electrons.

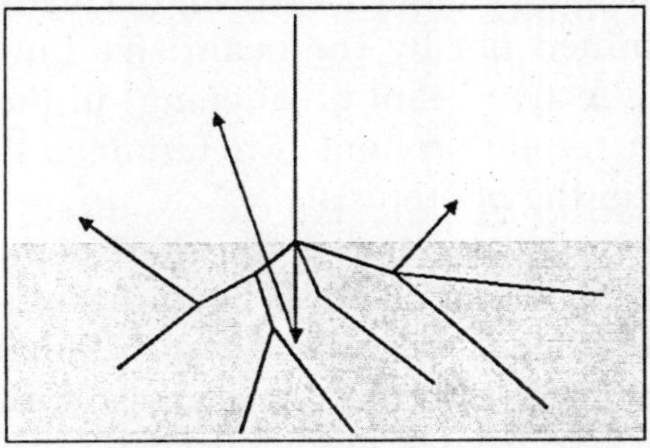

Fig. Electron Trajectories

In such a collision the momentum transfer from the incident electron to an atomic electron can be expressed as $dp = 2e^2/\ bv$, where b is the distance of closest approach between the electrons, and v is the incident electron velocity. The energy transferred by the collision is given by $T = (dp)^2/\ 2m = e^4/\ Eb^2$, where m is the electron mass and E is the incident electron energy, given by $E = (1/\ 2)mv^2$.

By integrating over all values of T between the lowest binding energy, Eo, and the incident energy, one obtains the result that the total cross section for collision is inversely proportional to the incident energy E, and proportional to $1/\ Eo$ "$1/\ E$. Generally, $E >> Eo$, so the result is essentially inversely proportional to the binding energy.

By using the same integration approach, but over the range $2Eo$ to E, one obtains by comparing cross-sections that half of the inelastic collisions of the incident electrons produce electrons with kinetic energy greater than Eo. These secondary electrons are capable of breaking bonds (with binding energy Eo) at some distance away from the original collision. Additionally,

They can generate additional, lower energy electrons, resulting in an electron cascade. Hence, it is important to recognize the significant contribution of secondary electrons to the spread of the energy deposition.

Resolution Capability

With today's electron optics, electron beam widths can routinely go down to a few nm. This is limited mainly by aberrations and space charge. However, the feature resolution limit is determined not by the beam size but by forward scattering (or effective beam broadening) in the photoresist while the pitch resolution limit is determined by secondary electron travel in the photoresist.

This point is driven home by the 2007 demonstration of double patterning using electron beam lithography in the fabrication of 15 nm half-pitch zone plates. Although a 15 nm feature was resolved, a 30 nm pitch was still difficult to do, due to secondary electrons scattering from the adjacent feature. The use of double patterning allowed the spacing between features to be wide enough for the secondary electron scattering to be significantly reduced.

The forward scattering can be decreased by using higher energy electrons or thinner photoresist, but the generation of secondary electrons is inevitable. The travel distance of secondary electrons is not a fundamentally derived physical value, but a statistical parameter often determined from many experiments or Monte Carlo simulations down to < 1 eV.

This is necessary since the energy distribution of secondary electrons peaks well below 10 eV. Hence, the resolution limit is not usually cited as a well-fixed number as with an optical diffraction-limited system. Repeatability and control at the practical resolution limit often require considerations not related to image formation, e. g., photoresist development and intermolecular forces.

Scattering

In addition to producing secondary electrons, primary electrons from the incident beam with sufficient energy to penetrate the photoresist can be multiply scattered over large distances from underlying films and/or the substrate. This leads to exposure of areas at a significant distance from the desired exposure location.

For thicker electrons, as the primary electrons move

forward, they have an increasing opportunity to scatter laterally from the beam-defined location.

This scattering is called forward scattering. Sometimes the primary electrons are scattered at angles exceeding 90 degrees, i. e., they no longer advance further into the resist. These electrons are called backscattered electrons and have the same effect as long-range flare in optical projection systems. A large enough dose of backscattered electrons can lead to complete exposure of resist over an area much larger than defined by the beam spot.

Proximity Effect

The smallest features produced by electron beam lithography have generally been isolated features, as nested features exacerbate the proximity effect, whereby electrons from exposure of an adjacent region spill over into the exposure of the currently written feature, effectively enlarging its image, and reducing its contrast, i. e., difference between maximum and minimum intensity. Hence, nested feature resolution is harder to control. For most resists, it is difficult to go below 25 nm lines and spaces, and a limit of 20 nm lines and spaces has been found.

Proximity effects (due to electron scattering) can be reduced by solving the inverse problem and calculating the exposure function $E(x,y)$ that leads to a dose distribution as close as possible to the desired dose $D(x,y)$ when convolved by the scattering distribution point spread function $PSF(x,y)$.

CHARGING

Since electrons are charged particles, they tend to charge the substrate negatively unless they can quickly gain access to a path to ground. For a high-energy beam incident on a silicon wafer, virtually all the electrons stop in the wafer where they can follow a path to ground. However, for a quartz substrate such as a photomask, the embedded electrons will take a much longer time to move to ground.

Often the negative charge acquired by a substrate can be compensated or even exceeded by a positive charge on the surface due to secondary electron emission into the vacuum. The presence of a thin conducting layer above or below the resist is generally of limited use for high energy (50 keV or more) electron beams, since most electrons pass through the layer into the substrate.

The charge dissipation layer is generally useful only around or below 10 keV, since the resist is thinner and most of the electrons either stop in the resist or close to the conducting layer.

The range of low-energy secondary electrons (the largest component of the free electron population in the resist-substrate system) which can contribute to charging is not a fixed number but can vary from 0 to as high as 50 nm. Hence, resist-substrate charging is not repeatable and is difficult to compensate consistently. Positive charging is more tolerable than negative charging, because the latter can deflect the electron beam away from the desired exposure location.

ELECTRON BEAM RESIST PERFORMANCE

A study performed at the Naval Research Laboratory indicated that low-energy (10-50 eV) electrons were able to damage ~30 nm thick PMMA films. The damage was manifest as a loss of material.

- For the popular electron-beam resist ZEP-520, a pitch resolution limit of 60 nm (30 nm lines and spaces), independent of thickness and beam energy, was found.
- A 20 nm resolution had also been demonstrated using a 3 nm 100 keV electron beam and PMMA resist. 20 nm unexposed gaps between exposed lines showed inadvertent exposure by secondary electrons.

NEW FRONTIERS IN ELECTRON-BEAM LITHOGRAPHY

To get around the secondary electron generation, it will be imperative to use low-energy electrons as the primary radiation to expose photoresist. Ideally, these electrons should have energies on the order of not much more than several eV in order to expose the photoresist without generating any secondary electrons, since they will not have sufficient excess energy. Such exposure has been demonstrated using a scanning tunneling microscope as the electron beam source. The data suggest that electrons with energies as low as 12 eV can penetrate 50 nm thick polymer photoresist.

The drawback to using low energy electrons is that it is hard to prevent spreading of the electron beam in the photoresist. Low energy electron optical systems are also hard to design for high resolution. Coulomb inter-electron repulsion always becomes more severe for lower electron energy.

Another alternative in electron-beam lithography is to use extremely high electron energies (at least 100 keV) to essentially "drill" or sputter the material. This phenomenon has been observed frequently in transmission electron microscopy.

However, this is a very inefficient process, due to the inefficient transfer of momentum from the electron beam to the material. As a result it is a slow process, requiring much longer exposure times than conventional electron beam lithography. Also high energy beams always bring up the concern of substrate damage.

Interference lithography using electron beams is another possible path for patterning arrays with nanometer-scale periods. A key advantage of using electrons over photons in interferometry is the much shorter wavelength for the same energy.

Despite the various intricacies and subtleties of electron beam lithography at different energies, it remains the most practical way to concentrate the most energy into the smallest area. There has been significant interest in the development of multiple electron beam approaches to lithography in order to increase throughput. This work has been supported by SEMATECH and start-up companies such as Multibeam Systems and Mapper.

However, the degree of parallelism required to be competitive would need to be very high (at least 10 million); this is far in excess of most scheduled demonstrations.

LITHOGRAPHY IN THE MEMS

PATTERN TRANSFER

Lithography in the MEMS context is typically the transfer of a pattern to a photosensitive material by selective exposure to a radiation source such as light. A photosensitive material is a material that experiences a change in its physical properties when exposed to a radiation source. If we selectively expose a photosensitive material to radiation (e. g. by masking some of the radiation) the pattern of the radiation on the material is transferred to the material exposed, as the properties of the exposed and unexposed regions differs.

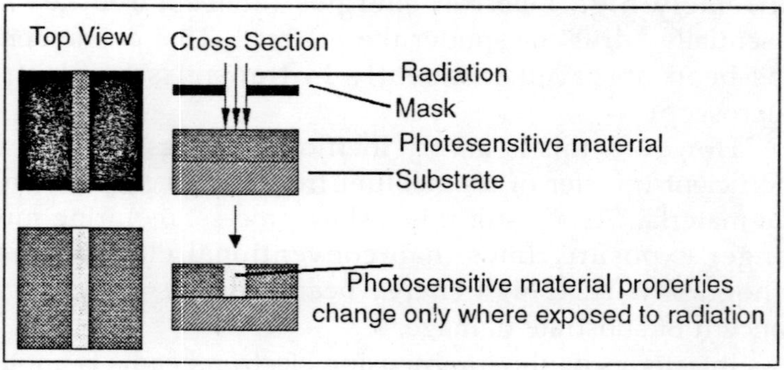

Fig. Transfer of a pattern to a photosensitive material

This discussion will focus on optical lithography, which is simply lithography using a radiation source with wavelength(s) in the visible spectrum.

In lithography for micromachining, the photosensitive material used is typically a photoresist (also called resist, other photosensitive polymers are also used). When resist is exposed to a radiation source of a specific a wavelength, the chemical resistance of the resist to developer solution changes. If the resist is placed in a developer solution after selective exposure to a light source, it will etch away one of the two regions (exposed or unexposed). If the exposed material is etched away by the developer and the unexposed region is resilient, the material is considered to be a positive resist. If the exposed material is resilient to the developer and the unexposed region is etched away, it is considered to be a negative resist.

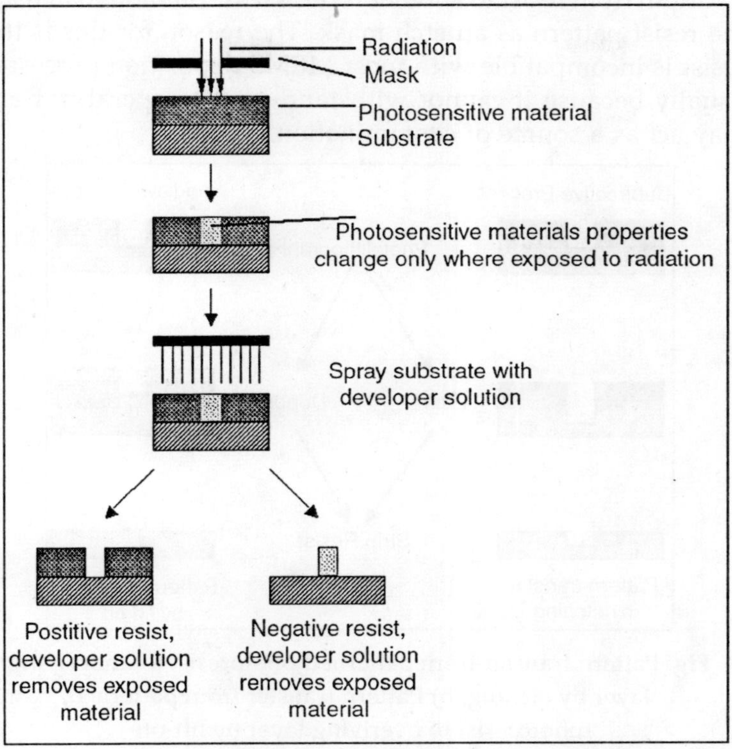

Fig. a) Pattern Definition in Positive Resist,
b) Pattern Definition in Negative Resist

Lithography is the principal mechanism for pattern definition in micromachining. Photosensitive compounds are primarily organic, and do not encompass the spectrum of materials properties of interest to micro-machinists. However, as the technique is capable of producing fine features in an economic fashion, a photosensitive layer is often used as a temporary mask when etching an underlying layer, so that the pattern may be transferred to the underlying layer. Photoresist may also be used as a template for patterning material deposited after lithography. The resist is subsequently etched away, and the material deposited on the resist is "lifted off".

The deposition template (lift-off) approach for transferring a pattern from resist to another layer is less common than using the resist pattern as an etch mask. The reason for this is that resist is incompatible with most MEMS deposition processes, usually because it cannot withstand high temperatures and may act as a source of contamination.

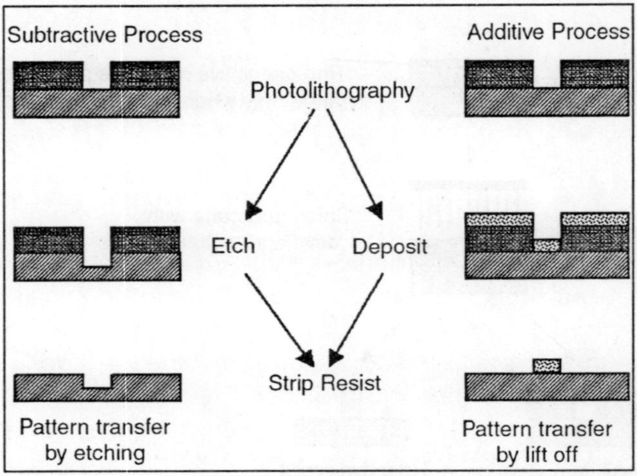

Fig. Pattern transfer from patterned photoresist to underlying layer by etching, b) Pattern transfer from patterned photoresist to overlying layer by lift-off

Once the pattern has been transferred to another layer, the resist is usually stripped. This is often necessary as the resist may be incompatible with further micromachining steps. It also makes the topography more dramatic, which may hamper further lithography steps.

ALIGNMENT

In order to make useful devices the patterns for different lithography steps that belong to a single structure must be aligned to one another.

The first pattern transferred to a wafer usually includes a set of alignment marks, which are high precision features that are used as the reference when positioning subsequent patterns, to the first pattern. Often alignment marks are included in other patterns, as the original alignment marks may be obliterated as processing progresses.

It is important for each alignment mark on the wafer to be labeled so it may be identified, and for each pattern to specify the alignment mark (and the location thereof) to which it should be aligned.

By providing the location of the alignment mark it is easy for the operator to locate the correct feature in a short time. Each pattern layer should have an alignment feature so that it may be registered to the rest of the layers.

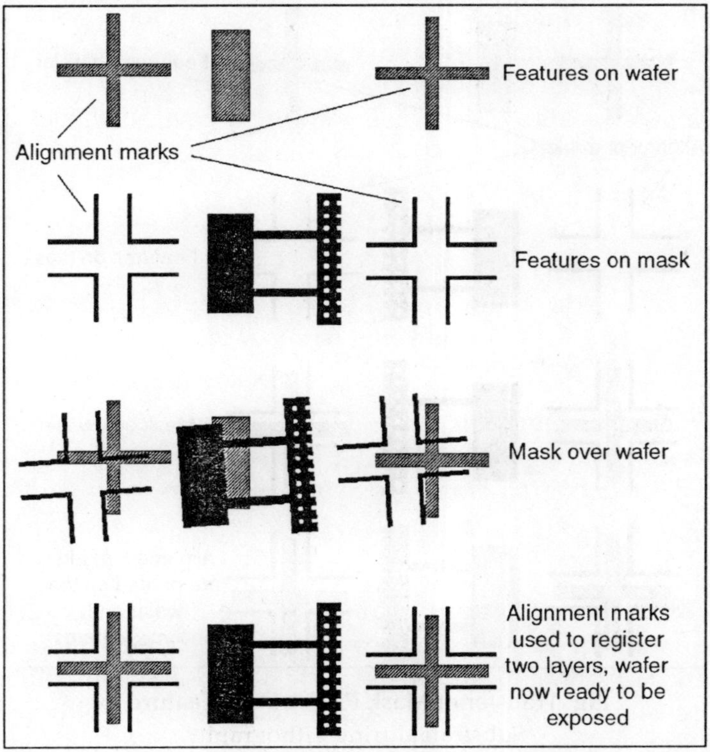

Fig. Use of Alignment Marks to Register Subsequent Layers

Depending on the lithography equipment used, the feature on the mask used for registration of the mask may be transferred to the wafer. In this case, it may be important to locate the alignment marks such that they don't effect subsequent wafer processing or device performance. For example, the alignment mark shown in figure will cease to exist after a through the wafer DRIE etch.

Pattern transfer of the mask alignment features to the wafer may obliterate the alignment features on the wafer. In this case the alignment marks should be designed to minimize this effect, or alternately there should be multiple copies of the alignment marks on the wafer, so there will be alignment marks remaining for other masks to be registered to.

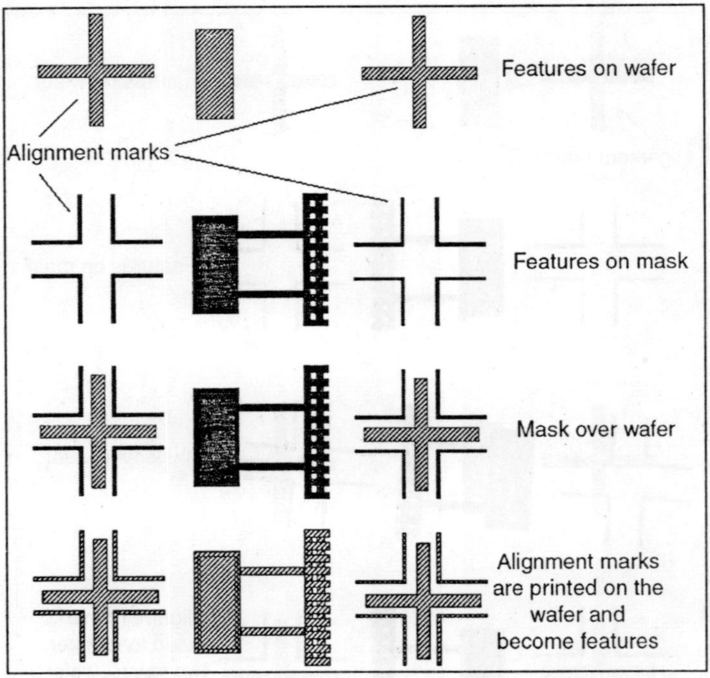

Fig. Transfer of Mask Registration Feature to Substrate during Lithography

Alignment marks may not necessarily be arbitrarily located on the wafer, as the equipment used to perform alignment may have limited travel and therefore only be able to align to features located within a certain region on the wafer. The region location geometry and size may also vary with the type of alignment, so the lithographic equipment and type of alignment to be used should be considered before locating alignment marks.

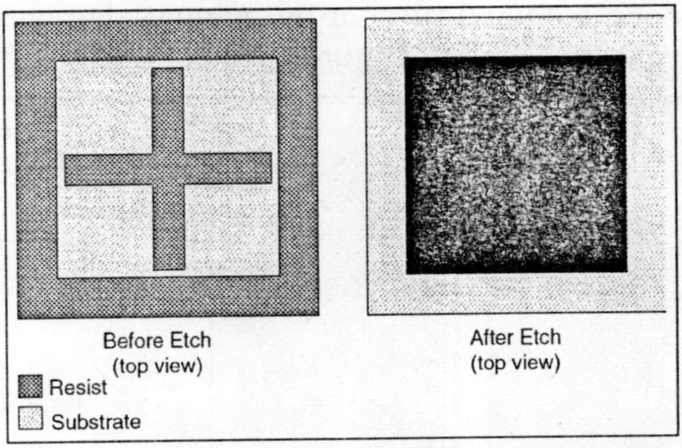

Fig. Poor Alignment Mark Design for a DRIE through the Wafer Etch

Typically two alignment marks are used to align the mask and wafer, one alignment mark is sufficient to align the mask and wafer in x and y, but it requires two marks (preferably spaced far apart) to correct for fine offset in rotation.

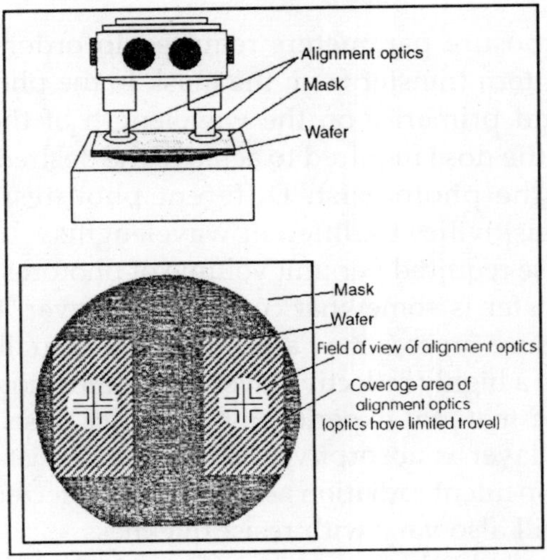

Fig. Restriction of Location of Alignment Marks Based on Equipment used

As there is no pattern on the wafer for the first pattern to align to, the first pattern is typically aligned to the primary wafer flat. Depending on the lithography equipment used, this may be done automatically, or by manual alignment to an explicit wafer registration feature on the mask.

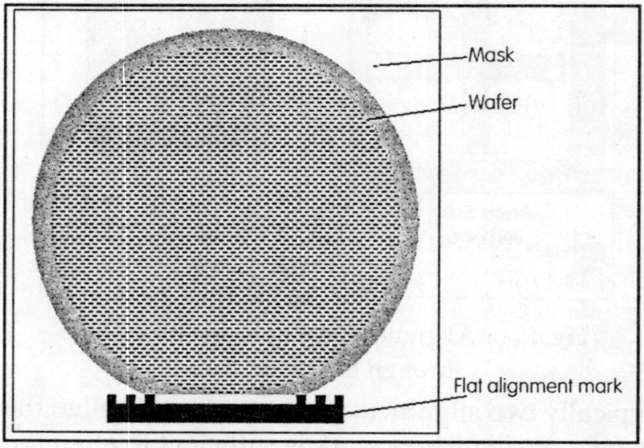

Fig. Mask Alignment to the Wafer Flat

Exposure

The exposure parameters required in order to achieve accurate pattern transfer from the mask to the photosensitive layer depend primarily on the wavelength of the radiation source and the dose required to achieve the desired properties change of the photoresist. Different photoresists exhibit different sensitivities to different wavelengths.

The dose required per unit volume of photoresist for good pattern transfer is somewhat constant; however, the physics of the exposure process may affect the dose actually received. For example a highly reflective layer under the photoresist may result in the material experiencing a higher dose than if the underlying layer is absorptive, as the photoresist is exposed both by the incident radiation as well as the reflected radiation. The dose will also vary with resist thickness.

There are also higher order effects, such as interference patterns in thick resist films on reflective substrates, which may affect the pattern transfer quality and sidewall properties. At the edges of pattern light is scattered and diffracted, so if an image is overexposed, the dose received by photoresist at the edge that shouldn't be exposed may become significant. If we are using positive photoresist, this will result in the photoresist image being eroded along the edges, resulting in a decrease in feature size and a loss of sharpness or corners.

If we are using a negative resist, the photoresist image is dilated, causing the features to be larger than desired, again accompanied by a loss of sharpness of corners. If an image is severely underexposed, the pattern may not be transferred at all, and in less sever cases the results will be similar to those for overexposure with the results reversed for the different polarities of resist.

If the surface being exposed is not flat, the high-resolution image of the mask on the wafer may be distorted by the loss of focus of the image across the varying topography. This is one of the limiting factors of MEMS lithography when high aspect ratio features are present. High aspect ratio features also experience problems with obtaining even resist thickness coating, which further degrades pattern transfer and complicates the associated processing.

THE LITHOGRAPHY MODULE

Typically lithography is performed as part of a well-characterized module, which includes the wafer surface preparation, photoresist deposition, alignment of the mask and wafer, exposure, develop and appropriate resist conditioning. The lithography process steps need to be characterized as a sequence in order to ensure that the remaining resist at the end of the modules is an optimal image of the mask, and has the desired sidewall profile.

The standard steps found in a lithography module are (in sequence): dehydration bake, HMDS prime, resist spin/spray, soft bake, alignment, exposure, post exposure bake, develop hard bake and descum. Not all lithography modules will contain all the process steps. A brief explanation of the process steps is included for completeness.

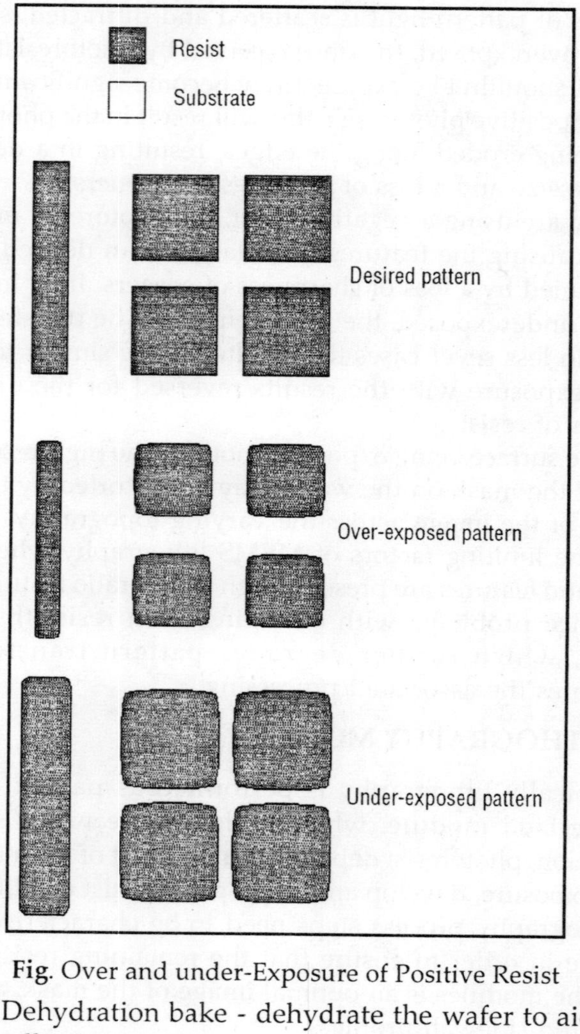

Fig. Over and under-Exposure of Positive Resist

- Dehydration bake - dehydrate the wafer to aid resist adhesion.
- HMDS prime - coating of wafer surface with adhesion promoter. Not necessary for all surfaces.
- Resist spin/spray - coating of the wafer with resist either by spinning or spraying. Typically desire a uniform coat.
- Soft bake - drive off some of the solvent in the resist,

may result in a significant loss of mass of resist (and thickness). Makes resist more viscous.

- Alignment - align pattern on mask to features on wafers.
- Exposure - projection of mask image on resist to cause selective chemical property change.
- Post exposure bake - baking of resist to drive off further solvent content. Makes resist more resistant to etchants (other than developer).
- Develop - selective removal of resist after exposure (exposed resist if resist is positive, unexposed resist if resist is positive). Usually a wet process (although dry processes exist).
- Hard bake - drive off most of the remaining solvent from the resist.
- Descum - removal of thin layer of resist scum that may occlude open regions in pattern, helps to open up corners.

We make a few assumptions about photolithography. Firstly, we assume that a well characterized module exists that: prepares the wafer surface, deposits the requisite resist thickness, aligns the mask perfectly, exposes the wafer with the optimal dosage, develops the resist under the optimal conditions, and bakes the resist for the appropriate times at the appropriate locations in the sequence.

Unfortunately, even if the module is executed perfectly, the properties of lithography are very feature and topography dependent. It is therefore necessary for the designer to be aware of certain limitations of lithography, as well as the information they should provide to the technician performing the lithography.

The designer influences the lithographic process through their selections of materials, topography and geometry. The material(s) upon which the resist is to be deposited is important, as it affects the resist adhesion. The reflectivity and roughness of the layer beneath the photoresist determines the amount of reflected and dispersed light present during exposure.

It is difficult to obtain a nice uniform resist coat across a surface with high topography, which complicates exposure and development as the resist has different thickness in different locations. If the surface of the wafer has many different height features, the limited depth of focus of most lithographic exposure tools will become an issue.

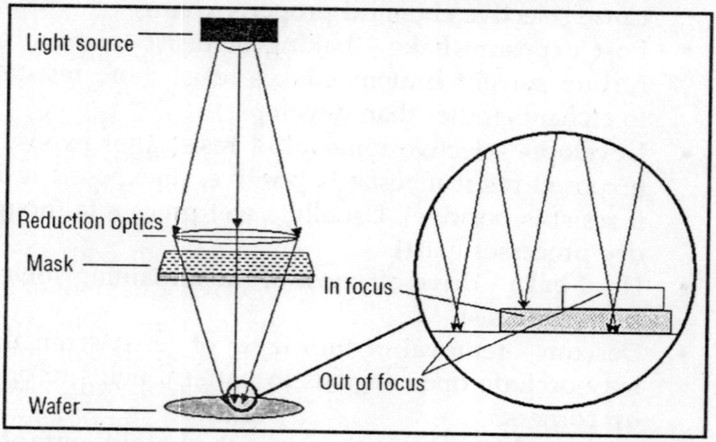

Fig. Lithography Tool Depth of Focus and Surface Topology

The designer should keep all these limitations in mind, and design accordingly. For example, it is judicious, when possible, to perform very high aspect patterning step (lithography and subsequent etch/deposition) last, as the topography generated often hampers any further lithography steps. It is also necessary for the designer to make it clear which focal plane is most important to them (keeping in mind that features further away in Z from the focal plane will experience the worst focus). The resolution test structures should be located at this level (as they will be used by the fab to check the quality of a photo step).

Chapter 6

Nanopattering

Lithography, based on traditional ink-printing techniques, is a process for patterning various layers, such as conductors, semiconductors, or dielectrics, on a surface. Nanopattering expands traditional lithographic techniques into the submicron scale.

A typical integrated circuit consists of various patterned thin films of metals, dielectrics and semiconductors on various substrates such as silicon, gallium arsenide, or germanium. Such a device is fabricated by a technology known as lithography, in which radiation sensitive polymeric materials called resists are used to produce circuit patterns in the substrates. Figure depicts the lithographic process sequence.

The resist material is applied as a thin coating, typically by spin coating over the substrate (wafer) and then heated to remove the casting solvent (post-apply bake, pre-exposure bake, or pre-bake). The resist film is subsequently exposed in an image-wise fashion through a mask (in photo- and X-ray lithography) or directly with finely focused electron beams. The exposed resist film is then developed typically by immersion in a developer solvent to generate three-dimensional relief images.

The exposure may render the resist film more soluble in the developer, thereby producing a positive-tone image of the mask.

Conversely, it may become less soluble upon exposure, resulting in generation of a negative-tone image. When the resist image is transferred into the substrate by etching and related processes, the resist film that remains after the development functions as a protective mask.

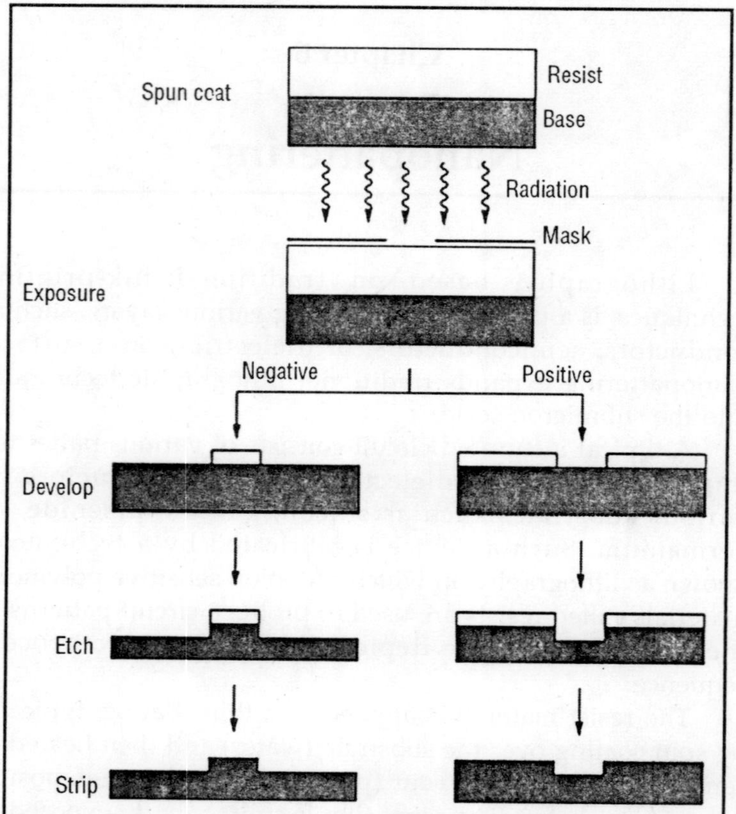

Fig. Schematic Representation of the Lithographic Process.

The resist film must "resist" the etchant and protect the underlying substrate while the bared areas are being etched. The remaining resist film is finally stripped, leaving an image of the desired circuit in the substrate. The process is repeated many times to fabricate complex semiconductor devices.

For a resist material to be useful in device fabrication:

- It must be capable of spin casting from solution into a thin and uniform film that adheres to various substrates such as metals, semiconductors, and insulators

- Possess high radiation sensitivity
- Possess high resolution capability, dictated by solubility/insolubility characteristics
- Withstand extremely harsh environments, for example, high temperature, strong corrosive acids, and plasmas such as used in subsequent etching, doping and sputtering operations.

In earlier generations of resists (often based on novolok phenolformaldehyde polymers), each absorbed quantum of radiation induced on average less than one chemical reaction within the material. "Photosolubilizable compositions" for photographic and photoresist applications with overall quantum yields greater than one were described by Smith as early as 1973. The idea was limited in practice, however, until thermostable yet photosensitive onium salts of strong acids were developed by Crivello as PAGs for cationic photopolymerization. The first CA resist for microlithography was reported by the university industry team of Fréchet (University of Ottawa) and Willson and Ito (IBM Inc.), who combined onium salts with the acid-deprotectable poly(4-[*t*-butyloxyca-rbonyloxy] styrene) (poly-TBOCST), and coined the term *chemical amplification*. Over 200 articles and several reviews have appeared on this topic between January 1992 and June 1994 alone.

In chemical amplification (CA) resists, the primary photochemical event produces a mobile catalyst that, typically during later postexposure baking (PEB), goes on to induce a cascade of material transforming secondary catalytic events within a 5-25 nm radius. Such chemical amplification thus makes possible an overall quantum yield (the number of material reactions divided by number of absorbed photons) of up to several hundred. Thus, a CA resist must contain:

- A small amount (ca. 1-5 wtper cent) of radiation-sensitive catalyst precursor, generally a photoacid generator (PAG);
- Many chemical groups that can react by elimination, addition, or rearrangement only in the presence of catalyst;

- A polymer matrix able to disperse all other components in a smooth clear film; and
- Optional additives to improve performance or processability; e. g., surfactants, photosensitizers, and etch resistors.

Although in most reported compositions the catalyst-sensitive groups are polymer-bound and the catalyst precursors are free, components i-iii or iv can in principle be interconnected in any combination, as small molecules, homopolymers, copolymers or blends.

Along with their higher sensitivity and contrast in forming images, CA are also better than earlier resists in being more flexible in design and formulation, versatile in radiation source (electromagnetic or particle beams), and compatible with dry (plasma), multilayer, and other advanced pattern transfer techniques. In general, resist systems can be classified on the basis of their design, namely, one-, two-, or multi-component systems.

One-component resist systems consist of pure polymers that must combine all the necessary attributes such as substrate protection, radiation sensitivity, and film-forming characteristics. The most popular resist designs in modern lithography are based on two-component systems in which resist functions are provided by two separate components.

Resist systems can also be divided into three groups on the basis of the radiation source: UV or photoresists, electron-beam resists, and X-ray resists. Photolithography that utilizes UV light has been the predominant technology in semiconductor manufacture and will continue to be so in the foreseeable future.

X-ray lithography is capable of producing high-resolution, high-aspect-ratio (height-width) images and is considered to be the technology of the future, whereas elctron-beam lithography is used in photomask fabrication. Photolithography can be further subdivided into near-UV (350-450 nm), mid-UV (300-350 nm), and deep-UV (< 300 nm) technologies, depending on the wavelength of the exposure.

The resolution is proportional to the exposing wavelength

and inversely proportional to the numerical aperture (NA) of the lens. Thus the i-line (365 nm) with a high-NA lens shifting from the g-line (436 nm), has been regarded as the dominant technology in the manufacture of 16-megabit (Mbit) dynamic random access memory (DRAM) devices with a minimum feature size of 0. 5 mm. Krypton fluoride (KrF, 248 nm), argon fluoride (ArF, 193 nm), and F2 (157nm) excimer laser technologies are emerging as the minimum feature size continues to shrink far below 0. 5 mm down to 80 nm.

The breakthrough that ultimately led to the adoption of 248 nm lithography as the technology of choice for advanced device fabrication was the development of CA resists. The consensus candidate for the next generation of optical lithography tools has been photolithograpy using 193 nm light. At this wavelength, the opacity of traditional aromatic-based materials precludes their use. Alternate resist materials based on aliphatic polymers and dissolution inhibitors have been examined.

The introduction of monomers bearing cyclic side groups such as adamantyl or norbornyl significantly improves the etch resistance over that of simple acrylic polymers. Alternately alicyclic structures have been incorporated directly into the polymer backbone offering a second route to 193 nm single layer resists.

The next generation of optical lithography at 157 nm has necessitated new developments in resist technology. Due to their high optical absorbance at 157 nm, single-layer 248 nm and 193 nm photoresists are not a viable alternative. The incorporation of fluorine into polymers have led the way to suitable resist materials.

NANOIMPRINT LITHOGRAPHY

Imprinting, or embossing, is a well-known technique to generate microstructures in hard polymers by pressing a rigid master containing surface-relief features into a thin thermoplastic polymer film that is then heated close to or, more generally, above the Tg. Nanoimprint lithography (NIL) has the potential of high-throughput due to the parallel processing, does not require sophisticated tools, and allows nanoscale replication for data storage.

NIL is also compatible with conventional device processing techniques. The quality of the nanoimprinting process depends on a number of experimental parameters like T, viscosity in the melt, adhesion of the polymer to the mold, etc. PMMA has been most widely used as the imprintable material, but a range of thermoplastic and thermosetting polymers is under investigation to optimize the imprinting and subsequent etching steps.

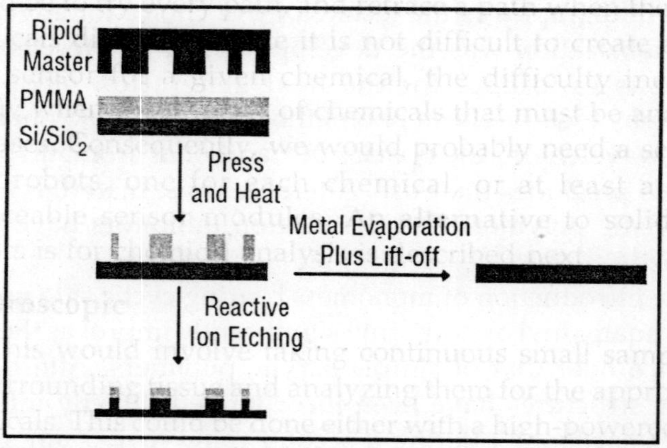

Fig. Schematic Overview of Nanoimprint Lithography.

The rigid master is usually prepared via e-beam lithography and has feature sizes in the 10–100 nm size range. After imprinting the polymer film, further etching can transfer the pattern into the underlying substrate. Alternatively, metal evaporation and lift-off of the polymer mask produces nanopattern metal features.

SOFT LITHOGRAPHY

Nanoimprint lithography (NIL) has primarily been used to emboss hard thermoplastic polymers. The micromolding and embossing of elastomers has attracted considerable interest as these materials have found important applications in softlithographic techniques such as microcontact printing (mCP). In this technique, a monolayer of a material is printed off an elastomeric stamp [made of poly(dimethylsiloxane) (PDMS)] after forming conformal contact between stamp and substrate.

Sub-micron surface relief structures can easily be introduced in PDMS by curing the polymers against a lithographically prepared master. The advantage of mCP is the ability to pattern surfaces chemically at the sub-micron level.

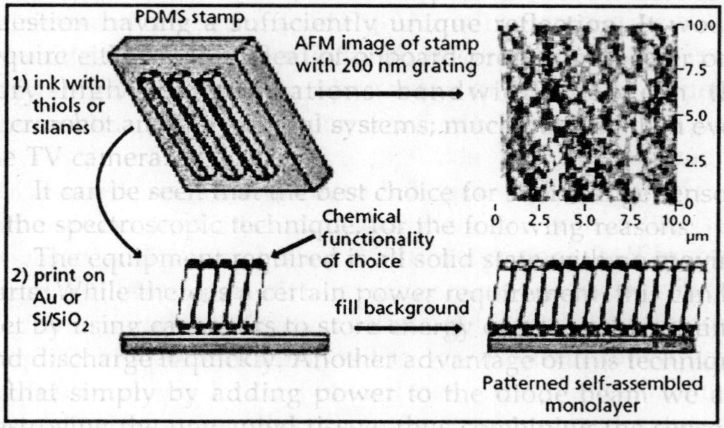

Fig. Schematic Overview of Microcontact Printing (mCP).

An elastomeric stamp is inked with small molecules (thiols or silanes) and pressed against a clean substrate (gold or silicon wafer). Where the stamp is in contact with the surface, a monolayer of material is transferred to the substrate. A second thiol or silane is used to fill in the background to provide a chemically patterned surface.

PHOTOCHEMICAL ACID GENERATORS

Photoacid generators (or PAGs) are cationic photoinitiators. A photoinitiator is a compound especially added to a formulation to convert absorbed light energy, UV or visible light, into chemical energy in the form of initiating species, viz., free radicals or cations.

Cationic photoinitiators are used extensively in optical lithography. The ability of some types of cationic photoinitiators to serve as latent photochemical sources of very strong protonic or Lewis acids is the basis for their use in photoimaging applications.

The continuing decrease in device dimensions in the microelectronics industry is being achieved by pushing the limits of optical lithography.

In chemically amplified resist technology, the radiation-sensitive material (resist) in which patterns are delineated typically includes a matrix polymer and an onium salt photoacid generator (or PAG). There are several materials' issues to be considered in the choice of the PAG: sufficient radiation sensitivity to ensure adequate acid generation for good resist sensitivity, absence of metallic elements, temperature stability, etc.

The usual photo-supplied catalyst has been strong acid. Triarylsulfonium and diaryliodonium salts have become the standard PAG ingredients in CA resist formulations, because of their generally easy synthesis, thermal stability, high quantum yield for acid (and also radical) generation, and the strength and nonvolatility of the acids they supply. Simple onium salts are directly sensitive to DUV, X-ray and electron radiations, and can be structurally tailored, or mixed with photosensitizers, to also perform well at mid-UV and longer wavelengths.

However, onium salts are ionic and many will phaseseparate from some apolar polymers, or not dissolve completely in some casting solvents. Nonionic PAGs such as phloroglucinyl and *o,o*-dinitrobenzyl sulfonates, benzylsulfones and some 1,1,1-trihalides are more compatible with hydrophobic media in general, although their thermal stabilities and quantum yields for acid generation are often lower.

The phenomenal rate of increase in the integration density of silicon chips has been sustained in large part by advances in optical lithography – the process, that patterns and guides the fabrication of the component semiconductor devices and circuitry.

Although the introduction of shorter-wavelength light

sources and resolution enhancement techniques should help maintain the current rate of device miniaturization for several more years, a point will be reached where optical lithography can no longer attain the required feature sizes.

Several alternative lithographic techniques under development have the capability to overcome these resolution limits – EUV, X-ray, electron beam and ion beam lithographies, but, at present, no obvious successor to optical lithography has emerged.

NANOPATTERNING

Techniques for fabricating on sub-micron length scales span a wide range, from sophisticated lithographic methods that have their origins in the semiconductor industry to more recent materials and chemical advances that rely on self-organization. For delineating patterns below 100 nm, several approaches have been proposed (and indeed demonstrated).

These include nano-imprint lithography (including micro-contact printing, mold-assisted lithography, and hot embossing lithography), near-field optical lithography, direct patterning on a nanometer scale with scanning-probe microscopes, self-assembly of monolayers, pattern formation based on phase separation of polymers, etc.

The search is on for nonphotolithographic methods that could provide technologically simpler and cheaper nanofabrication strategies. Some of these approaches are better suited for producing individual nano-structures for the investigation of nanometer-scale devices; the throughput is likely to always remain impracticably low for commercial application.

Others such as nanoimprint lithography have the potential of high throughput due to parallel processing, do not require sophisticated tools, and allow nanoscale replication for data storage.

The natural length scales of polymer chains and their morphologies in the bulk, which lie in the nanometer domain, make polymers ideal building blocks for nanopatterning. Recent developments in the use of polymers for the fabrication of nanostructures via lithographic and self-assembling strategies have been reviewed.

NANOPATTERNING VIA PHASE SEPARATION OF POLYMERS

Block copolymers of flexible, chemically incompatible, and dissimilar blocks can microphase separate into a variety of morphologies with nanometer scale dimensions. This self-assembly process is driven by an unfavorable mixing enthalpy and a small mixing entropy, while the covalent bond between the two blocks prevents macrophase separation.

When the microphase separated morphology can be controlled and turned into a useful structure, phase separation of block copolymers can be a powerful tool for fabricating nanostructures without additional lithography and processing steps.

In addition, to block copolymers copolymers, random copolymers comprising sufficiently dissimilar components based on size and chemical nature, for e. g., SSQ-MMA copolymer, SSQ/Polypropylene blend, or physical blends of hydrophobic and hydrophilic polymers driven by an underlying pattern of incompatibility have been shown to yield functional and arbitrary patterns. The use of surfactants in such blends may be used to accentuate the areas of different surface tension.

Soft lithographic approaches have been combined with surfactant and particulate templating procedures to create to create hierarchically ordered oxides. A recent report combines molecular scale, evaporation-induced self-assembly (EISA) of organically modified mesophases with macroscopic, evaporative printing procedures. This allows the rapid fabrication of hierarchical structures exhibiting form and function on multiple length scales and at multiple locations. The formulation of the "ink" in such hierarchial assembly strategies is complex and may use a variety of surfactants.

SELF-ASSEMBLED MONOLAYER SYSTEMS (SAMS)

Polyelectrolytes are defined as materials for which the

solution properties in solvents of high dielectric constant are governed by electrostatic interactions over distances larger than typical molecular dimensions.

These materials are widely used in industrial applications as dispersants in aqueous media, flocculating agents to coagulate slurries and industrial wastes, for sizing in textile and paper manufacture, and as conditioning additives to drilling muds and soil to prevent abrasive damage.

More recently, they have been applied in molecular self-assembly techniques for thin film deposition of electrically conducting polymers, conjugated polymers for light emitting devices, nanoparticles, and noncentrosymmetric-ordered second order nonlinear optical (NLO) devices.

The technique of Self-Assembled Monolayers or SAMs is an ingeniously simple, yet powerful nanoscale approach for the fabrication of functional supramolecular assemblies for various device applications. It involves the alternate adsorption of anionic and cationic polyelectrolytes onto a suitable substrate. Typically, only one of these is the active layer, while the other enables the composite multilayered film to be bound by electrostatic attraction.

Alternatively, the oppositely charged polyelectrolytes may serve as a barrier for the sustained release of an active core. Controlled formation of highly ordered, *three-dimensional*, multifunctional, reactive, thin films containing biological molecules is seeing widespread application in the areas of biotechnology and biomaterials science.

Of the potential polyanionic candidates, poly(styrenesulfonic acid) (PSSA) and its salts have been used extensively. Why? Excellent adsorption properties, water solubility, smooth films with easily controllable thickness, controlled level of loading/penetration of active component—all properties essential in realizing high performance devices based on supramolecular structures in terms of selectivity, sensitivity, response time, and stability.

In addition, PSSA has been used to dope thiophene-based conjugated polymers to make them conducting, e. g., PEDOT/PSS. Post treatment of polyaniline (emeraldine salt), grafted to lignin, with PSSA may also be used to enhance electrical conductivity.

NORBORNADIENE

A scan of the scientific and patent literature reveals that this versatile monomer has been applied over a vast spectrum of high technology applications, in fields ranging from materials science, ag-related, and pharmaceutical, to being used as a model system in fundamental research activity (for example, in testing new nanocatalysts - in single-step hydrogenations, tandem cycloaddition reactions, Pauson-Khand annelation as well as in comprehensive theoretical studies.

The unique set of properties offered by Norbornadiene along with its transformations driving some of the reported applications are highlighted in the following table.

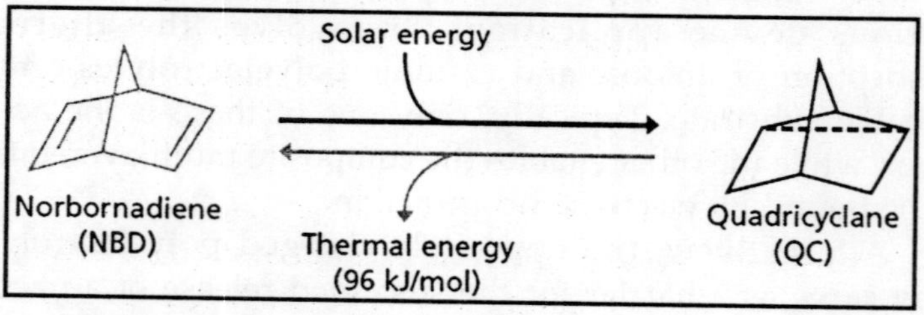

Scheme: Conversion Between Norbornadiene (NBD) and Quadricylane (QC).

SCIENCE OF NANOSCALE STRUCTURES

This science of nanoscale structures deals with the creation, investigation and utilisation of systems that are 1000 times smaller than the components currently used in the field of microelectronics. Biotechnology deals with metabolic process with microoraganisms. Convergence of these two technologies results in growth of nanobiotechnology. This interdisciplinary combination can create many innovative tools.

The biomedical applications of nanotechnology are the direct products of such convergences.

Table: Properties of Norbornadiene

Bicyclic, strained system	Undergoes photochemical valence isomerization to Quadricyclane (QC), converting solar energy to strain energy	Solar energy storage system8
Bicyclic, strained system	Strain energy released gradually as heat upon conversion back to NBD	Microheater; Energetic binder for solid rocket propellants10
Diolefin	Starter for norbornenediol derivative to synthesize stereo-controlled ROMP-derived precursor polymer, which upon thermal elimination results in a special form of polyacetylene	Synthon for Conductive Polymer, polyacetylene
Cycloaliphatic Monomer	Controlled cyclopolymerization of bulky ester derivatives of NBD to afford regioregular 2-alkoxycarbonyl nortricyclene polymers	New resist materials based on cycloaliphatic polymers for imaging with 193 nm lasers
Rigid, 3-D, crosslinking agent	Synthesis of soybean oilstyrene-NBD thermosetting copolymers	Shape memory polymer with Tg well above ambient
NBD	ROMP synthesized fluoropolymer, poly[2,3-bis (trifluoromethyl)NBD] having usefully high values of pyroelectric properties combined with low dielectric loss at RT	Pyroelectric transducer with a figure of merit comparable or better than that of PVDF
NBD	Controlled selectivity and reaction rate for cyclic dimerization, co-dimerization,	As a universal substrate for organic and petrochemical synthesis of wide

	isomerization, and allylation of NBD	range of rare polycyclic hydrocarbons
NBD	Inhibited ethylene production and RNAase induction in apple pulp tissue	Growth regulator in fruit production
Highly strained NBD	Reagent in the Pd-catalyzed conversion of hydrosilanes to alkoxysilanes	Efficient hydrogen scavenger
Bicyclic NB	Starter in a fifteen step synthesis of (+)-Sparteine	Synthon in the first asymmetric total synthesis of (+)-Sparteine

However, the challenges facing scientists and engineers working in the field of nanotechnology are quite enormous and extraordinarily complex in nature.

Utility of nanotechnology to biomedical sciences imply creation of materials and devices designed to interact with the body at sub-cellular scales with a high degree of specificity. This could be potentially translated into targeted cellular and tissue-specific clinical applications aimed at maximal therapeutic effects with very limited adverse-effects.

Nanotechnology in biomedical sciences presents many revolutionary opportunities in the fight against all kinds of cancer, cardiac and neurodegenerative disorders, infection and other diseases.

This article presents an overview of some of the applications of nanotechnology in biomedical sciences.

NEW AREA OF SCIENCE

Nanotechnology is a new area of science that involves working with materials and devices that are at the nanoscale level. A nanometre is billionth of a meter. That is, about 1/80,000 of the diameter of a human hair, or ten times the diameter of a hydrogen atom. It manipulates the chemical and physical properties of a substance on molecular level. Nanotechnology alters the way we think, it blurs the boundaries between physics, chemistry and biology, the elimination of these boundaries will pose many challenges and new directions for the organisation of education and research.

Richard Feynman's speech called 'There is plenty of room at the bottom' in 1959 emphasised this concept - If our small minds, for some convenience, divide this universe into parts, physics, biology, geology, astronomy, psychology and so on.

Nanobiotechnology is the unification of biotechnology and nanotechnology. This hybrid discipline can also mean making atomic-scale machines by imitating or incorporating biological systems at the molecular level, or building tiny tools to study or change natural structure properties atom by atom. Nanobiotechnology can have a combination of the classical micro-technology with a molecular biological approach. Biotechnology uses the knowledge and techniques of biology to manipulate molecular, genetic, and cellular processes to develop products and services, and is used in diverse fields from medicine to agriculture.

Convergence, is an activity or trend that occurs based on common materials and capabilities-in this case the discipline that enables convergence is nanotechnology. The potential opportunities offered by this interface is truly outstanding; the overlap of biotech, nanotech and information technology is bringing to fruition many important applications in life sciences.

This technology is expected to create innovations and play a vital role in various biomedical applications, not only in drug delivery and gene therapy, but also in molecular imaging, biomarkers and biosensors. Target-specific drug therapy and methods for early diagnosis of pathologies are the priority research areas where nanotechnology would play a prominent role.

The National Institutes of Health Bioengineering Consortium (BECON) held a symposium in 2000 entitled "Nanoscience and Technology: Shaping Biomedical Research3. Eight areas of nanoscience and nanotechnology were addressed at the conference and believed to be the most pertinent to research in biomedicine. These areas included synthesis and use of nanostructures, applications of nanotechnology to therapy, biomimetic and biologic nanostructures, electronic-biology interface, devices for early detection of disease, tools for the study of single molecules, nanotechnology and tissue engineering.

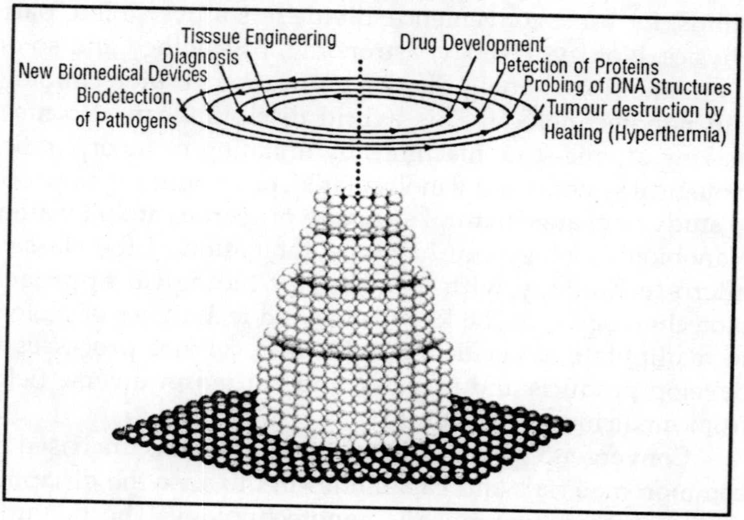

Fig. Schematic Illustration of Nanotechnology Revolutionising Biomedical Sciences

The aim of BECON was to enhance communication between biomedical scientists and engineers who bring different aspects of their skills and knowledge to bear on these problems and to make the biomedical community more aware of the emerging developments in the field of nanotechnology. The deliberations of the conference are now widely reinforced by day-to-day experience, increasing ability to manipulate individual molecules at a nanoscale and to combine biomolecules with other nanoscale structures. This ability provides the opportunity for untold new therapeutic and diagnostic applications by enabling the building of novel structures from the bottom up.

In the foreseeable future, the most important clinical

application of nanotechnology will probably be in pharmaceutical development. These applications take advantage of the unique properties of nanoparticles as drugs or constituents of drugs or are designed for new strategies to controlled release, drug targeting, and salvage of drugs with low bioavailability.

Nanoscale polymer capsules can be designed to break down and release drugs at controlled rates, to allow differential release in certain environments, such as an acid medium, and to promote uptake in tumours versus normal tissues. A lot of research is now focused on creating novel polymers and exploring specific drug-polymer combinations.

Nanocapsules can be synthesized directly from monomers or by means of nanodeposition of preformed polymers. Nanocapsules have also been formulated from albumin and liposomes. Implantable drug delivery systems that are being developed will make use of nanopores to control drug release.

One of the key issues in bio-availability is cell transfection in DNA gene therapy. Current methods have significant limitations, including the risk of inadvertent transmission of disease by viral vectors. This has led researchers to explore polymer-DNA complexes and liposome-DNA complexes for gene delivery. It has also been shown that compacted DNA in the form of nanoparticles can be used to transfect postmitotic cells.

Despite the risk and limitations, viral vectors are an efficient biomimetic approach to drug targeting and delivery. The tat peptide from human immunodeficiency virus (HIV) and other viral proteins are being attached to DNA, proteins, and other materials for uptake into cells. These nano-assemblies mimic the action of the fusion proteins that make viral transfection efficient.

Nanotechnology has also enabled the development of biochips and has a role in green manufacturing (e. g biocompatibility and biocomplexity areas). Other applications include the design of sensors for astronauts, soldiers, biofluids (for handling DNA and other molecules), in vitro fertilization of live stock, nanofiltration, bioprocessing 'by design' and traceability of genetically modified food. Exploratory areas for nanotechnology will include research into the condition and/ or repair of the brain and other areas for regaining cognition.

It might also find application in designing pharmaceuticals as a function of patient genotypes and in applying chemicals to stimulate production as a function of plant genotypes. The synthesis of more effective and biodegradable chemicals for agriculture and the production of implantable detectors could be aided by nanotechnology with minimal quantities of blood.

Employing this technology it should also be possible to develop methods that use saliva instead of blood for the detection of illnesses or that can perform complete blood testing within a short period of time. Broader issues include economic molecular medicine, sustainable agriculture, conservation of biocomplexity, and enabling emerging technologies.

Richard E. Smalley, winner of the 1996 Nobel Prize in Chemistry announced in his congressional testimony to the U. S. government about the increasing awareness in the scientific and technical community of our entry into a new golden age. Burgeoning interest in the medical applications of nanotechnology has led to the emergence of a new discipline known as nanomedicine.

On a wider scope, nanomedicine is the process of diagnosing, treating, preventing disease and traumatic injury, of relieving pain, and of preserving and enhancing human health, using molecular tools and molecular knowledge of the human body. The purpose of this review is to throw more light on the recent advances and impact of nanotechnology on biomedical sciences.

RECENT DEVELOPMENTS

Medical diagnosis with appropriate and effective delivery of pharmaceuticals are the medical areas where nanosize particles have found practical applications. However, there are many other interesting proposals for the use of nanomechanical tools in the fields of medical research and clinical practice. Such nanotools are awaiting construction, and presently are more like a fantasy. Nevertheless, they might be quite useful, and become a reality in the near future.

Nanodevices in medical sciences could function to replace defective or improperly functioning cells, such as the respirocyte proposed by Freitas. This man-made red blood cell is theoretically capable of providing oxygen more effectively than an erythrocyte.

It could replace defective natural red cells in blood circulation. Primary applications of respirocytes may involve transfusable blood substitution, partial treatment of anaemia, prenatal/neonatal problems, and lung disorders.

It has been reported that nanomachines could administer drugs within a patient's body. Such nanoconstructions could deliver drugs to peculiar sites making treatment more accurate and precise. Similar machines with specific 'weapons' could be used to remove obstacles in the circulatory system or in the identification and killing of tumour cells.

The other vital application of nanotechnology in relation to medical research and diagnostics are nanorobots. Nanorobots, operating in the human body, could monitor levels of different compounds and record the information in the internal memory.

They could be rapidly used in the examination of a given tissue, surveying its biochemical, biomechanical, and histometrical features in greater detail. Just as biotechnology extends the range and efficacy of treatment options available from nanomaterials, the advent of molecular nanotechnology will again expand enormously the effectiveness, comfort and speed of future medical treatments while at the same time significantly reducing their risk, cost, and invasiveness.

Biotechnology permits tailor-made production and biopharmaceuticals and biotechnological drugs, many of which require special formulation technologies to overcome drug-associated problems.

Such major challenges to solve include the following: poor solubility, limited chemical stability in vitro and in vivo after administration (i. e. short half-life), poor bioavailability and potentially strong side-effects requiring drug enrichment at the site of action (targeting).

Nanoparticulate carriers have been developed as one solution to overcome such delivery problems, i. e. drug nanocrystals, solid lipid nanoparticles (SLN), nanostructured lipid carriers (NLC) and lipid-drug conjugate (LDC) nanoparticles. The carriers as reported by Muller and colleagues are suitable to solve delivery problems with biotech drugs of different solubility.

Targeting with these carriers can be realised by a very simple approach, the differential protein adsorption. This technology proved to be efficient enough to accumulate sufficiently high amounts of drugs in the brain to reach therapeutic levels and also fulfill the major requirement to be pursued by a pharmaceutical company.

Quantum Dot with nanodots of a specific colour are believed to be flexible and could offer a cheap and easy way to screen a blood sample for the presence of a number of different viruses at the same time. It could also give physicians a fast diagnosis tool to detect, say, the presence of a particular set of proteins that strongly indicates the onset of myocardial infarction.

On the research front, the ability to simultaneously tag multiple biomolecules both on and inside cells could allow scientists to watch the complex cellular changes and events associated with disease, providing valuable clues for the development of future pharmaceuticals and therapeutics.

The National Heart, Lung, and Blood Institute (NHLBI) plans to foster the application of nanotechnology to HLBS (Heart, Lung, Blood and Sleep) research and disorders. A request for information (RFI) was developed, with advice from scientists and physicians with interests in nanotechnology, to canvas the broader scientific community on approaches to developing and applying nanotechnology to HLBS disorders.

A working Group comprising scientists, engineers, and physicians with expertise across nanotechnology, nanoscience, and HLBS medicine met on February 28th, 2003, using the RFI responses as the starting point for discussions.

The Working Group was entrusted with assessing the field of nanotechnology and suggesting ways for research. The Working Group cautioned against overly rigid or restrictive definition of nanotechnology, emphasizing the continuum of scale from the nanoscale to the microscale.

The Group also identified areas of opportunity and challenges to further development associated with the application of nanoscience and nanotechnology to improved diagnosis, treatment, and prevention of HLBS disorders. It as well developed prioritized recommendations to facilitate the application of nanotechnology to biological questions and improved patient care.

The RESIST Group at the Welsh School of Pharmacy at Cardiff University and others have looked at how molecularly imprinted polymers could be medically useful in clinical applications such as controlled drug release, drug monitoring devices, and biological and antibody receptor mimics. Histamine and ephedrine molecularly imprinted polymers (MIPs) were studied as potential biological receptor mimics whilst a propanolol MIP was investigated for its use as a rate attenuating selective excipient in a transdermal controlled device.

The first artificial voltage-gated molecular nanosieve was fabricated by Charles R. Martin and colleagues [48] at Colorado State University in 1995. Martin's membrane contains an array of cylindrical gold nanotubules with inside diameters as small as 1. 6nm. When the tubules are positively charged, positive ions are excluded and only negative ions are transported through the membrane.

When the membrane receives a negative voltage, only positive ions can pass. Similar nanodevices may combine voltage gating with pore size, shape, and charge constraints to achieve precise control of ion transport with significant molecular specificity. An exquisitely sensitive ion channel switch biosensor was built by an Australian research group.

The year 2003 could be termed a very special year for biomedical research because we celebrated the completion of the sequencing of the entire human genome which coincided with the 50th anniversary of the discovery of the DNA double helix structure by Watson and Crick. In biomedical imaging, we also witnessed the awarding of the Nobel Prize in Medicine and Physiology to two pioneers in Magnetic Resonance Imaging, Professor Paul Lauterbur and Sir Peter Mansfield.

These landmark events helped to highlight the impact of the rapid development in many diverse disciplines to biomedical research. The leverage and tremendous advances in electronics and information technology has been brought about by biomedical imaging research.

The opportunities and challenges in future biomedical research lie in the incorporation of knowledge gained from molecular biology with chemistry, physics, engineering, information technology, and nanotechnology to understand the ambiguity and complexity of life and come up with new diagnostic and therapeutic methods.

Calcium phosphate nanoparticles present a unique class of non-viral vectors, which can serve as efficient and alternative DNA carriers for targeted delivery of genes. The design and synthesis of ultra-low size, highly monodispersed DNA doped calcium phosphate nanoparticles of size around 80nm in diameter has been reported. The DNA encapsulated inside the nanoparticle is protected from the external DNase environment and could be used safely to transfer the encapsulated DNA under in vitro and in vivo conditions.

The application of a combination of nanomedicine with biophotonics for optically tracking the cellular pathways of gene delivery and the resulting transfection by using nanoparticles as a non-viral vector has been demonstrated recently. Gene delivery is an area of considerable current interest; genetic materials (DNA, RNA, and oligonucleotides) have been used as molecular medicine and are delivered to specific cell types to either inhibit some undesirable gene expression or express therapeutic proteins.

NANO-DNA TECHNOLOGY

The discovery of the polymerase chain reaction (PCR) paved the way to a new era of biological research. The impact can be felt not only in the field of molecular biology, but also in other allied fields of science. Novel classes of semi-synthetic DNA-protein conjugates, self-assembled oligomeric networks consisting of streptavidin and double-stranded DNA, which can be converted into well-defined supramolecular nanocircles have been developed.

The DNA-streptavidin conjugates are applicable as modular building blocks for the production of new immunological reagents for the ultrasensitive trace analysis of proteins and other antigens by means of immuno-PCR methodology. Immuno-PCR is a combination of the specificity of an antibody-based immuno-assay with the exponential power of the amplification of PCR, hence resulting in a 1000-fold degree of sensitivity as compared with standard ELISA (Enzyme-linked immunosorbent assay) methods.

Self-assembled DNA-streptavidin conjugates have also been applied in the field of nanotechnology. For example, the conjugates are used as model systems for ion-switchable nanoparticle networks, as nanometre-scale 'soft material' calibration standards for scanning probe microscopy, or as programmed building blocks for the rational construction of complex biomolecular architecture, which may be used as templates for the growth of nanometre-scale inorganic devices.

Covalent conjugates of single-stranded DNA and streptavidin are used as biomolecular adapters for the immobilization of biotinylated macromolecules at solid substrates through nucleic acid hybridization.

This 'DNA-directed immobilization' allows for reversible and site-selective functionalization of solid substrates with metal and semiconductor nanoparticles or, vice versa, for the DNA directed functionalization of gold nanoparticles with proteins, such as immunoglobulins and enzymes. The fabrication of functional biometallic nanostructures from gold nanoparticles and antibodies are applied as diagnostic tools in bioanalytics.

Nanobiotechnology in Nucleotide Polymorphism Analysis

The human genome sequence containing over two million

single nucleotide polymorphisms (SNPs), the next challenge is the development of the technologies to use this information in a cost-effective manner. Genotyping methods have to be improved in order to increase throughput by at least two orders of magnitude to enable pharmaceutical, biotechnological and academic research to uncover associations between genetic variants and diseases, with consequent potential for the development of novel diagnostics and therapies.

New approaches to DNA extraction and amplification have curtailed the times required for these processes to seconds. Microfluidic devices enable polymorphism detection through very rapid fragment separation using capillary electrophoresis and high-performance liquid chromatography, together with mixing and transport of reagents and biomolecules in integrated systems. The basic objectives in the development of a DNA extraction and purification system that will be compatible with high-throughput SNP genotyping requirements are:

- Release of the DNA from the cells into solution without either enzymatic (i. e. endonucleases) or mechanical (shearing) breakdown of the DNA;
- Removal of cellular debris (e. g. proteins) that may hamper DNA amplification or hybridization assays;
- High-throughput and economical DNA sample preparation with simplified protocols that reduce the number of procedures involved;
- Avoidance of hazardous chemical requirements as much as possible to minimize handling and disposal costs;
- Consistency of both quality and quantity of DNA yield among samples so that quantification is unnecessary, and subsequent amplification and/or hybridization can be to a high degree of reproducibility;
- A highly efficient process, to ensure enough supply for the enormous number of assays anticipated; and
- An interface that will enable direct loading of conventionally sampled biopsies on to the system.

The potential for nanotechnology to contribute to rapid high-throughput SNP analysis is most evident with smart biochip platforms. The development of an electronically addressable microarray platform as described by Heller L. et al 2000 has given rise to Nanogen Inc. The challenge of providing one or more technology platforms capable of SNP screening throughput of the order of 10^7 genotypes per day will need to be achieved, to allow significant associations between genes and diseases to be established.

Additionally, the technology platform(s) will also need to deliver economies of scale, such that the cost per genotype will be less than 0. 01$ for the magnitude of screening necessary to be feasible. From the rapidly developing field of nanotechnology, novel tools and processes have been introduced with the potential to provide the capabilities required.

Differences of SNPs occurring in close proximity to each other on the genome is normally correlated due to linkage during the process of replication, and the extent of this correlation is termed linkage disequilibrium. Where a significant association occurs between the genetic variation observed at specific SNPs and the presence of a disease, susceptible genes can be identified.

The statistical estimations needed to eliminate false-positive results were reviewed by McCarthy and Hilfiker. They suggest a linear increase in sample size is necessary for every order of magnitude increase in the number of markers tested. Hence, positive identification of a susceptible gene from a screening programme including 1 Million SNPs would require a minimum sample size of 1000 (i. e. a minimum of 10^9 SNPs have to be screened).

NANOPARTICLES AS BIOMARKERS

Nanoparticles can be used for both quantitative and qualitative in vitro detection of tumour cells. They enhance the detection process by concentrating and protecting a marker from degradation, in order to render the analysis more sensitive. For instance, streptavidin-coated fluorescent polystyrene nanospheres Fluospheres (green fluorescence) and TransFluospheres (red fluorescence) were applied in single colour flow cytometry to detect the epidermal growth factor receptor (EGFR) on A431 cells (human epidermoid carcinoma cells). The results have shown that the fluorescent nanospheres provided a sensitivity of 25 times more than that of the conjugate streptavidin-fluorescein.

New tools can now be developed, designed at the intersection of proteomics and nanotechnology, whereby nanoharvesting agents can be instilled into the circulation (e. g. derivatized gold particles) or into the blood collection devices to act as 3 molecular mops3 that soak up and amplify the bound and complexed biomarkers that exist.

These nanoparticles, with their bound diagnostic cargo, can be directly queried via mass spectrometry to reveal the low molecular weight and enriched biomarker signatures. Ultimately, utility of any approach for detecting disease is assessed on its clinical impact to patient outcome and disease-free survival. What is urgently required in the study of diseases in general, is the development of biomarkers that can detect curable diseases earlier, and not detecting advanced disease better.

Contrast agents have been loaded onto nanoparticles for tumour diagnosis purposes. The physico-chemical features (particle size, surface charge, surface coating, stability) of the nanoparticles allow the redirection and the concentration of the marker at the specific site of interest. Labelled colloidal particles could be used as radiodiagnostic agents. On the other hand, some non-labelled colloidal systems are already in use and some are still being tested as contrast agents in related diagnosis procedures such as computed tomography and NMR imaging.

To date, a study of radionucleide use in diagnostic imaging with nanoparticles for cancer detection is yet to be published. However, as conventional colloidal particles can be cells of organs like the liver, the spleen, the lungs and the bone marrow and as long-circulating nanoparticles can have a compartmental localization in the blood circulation or the lymphatic system- all these organs being potential sites for tumour development, these colloidal systems could potentially improve tumour diagnosis.

In the future, nanoparticles that are engineered with specific binding affinities can be resuspended into the collected body fluids, or perhaps even injected directly into the circulation. The nanoparticles, together with the bound molecules, could be directly captured on engineered filters and directly questioned by ultra high-resolution mass spectrometry (e. g. Fourier Transform Ion Cyclotron Resonance).

Nanotechnology in Measurements of Dissolved Oxygen

Oxygen is one of the major metabolites in aerobic systems, and the measurement of dissolved oxygen is of vital importance in medical, industrial, and environmental applications. Recent interest in the methods for measuring dissolved oxygen concentration has been focused mainly on optical sensors, due to their advantages over conventional amperometric electrodes in that they are faster, do not consume oxygen, and are not easily poisoned.

Optical Pebble (probes encapsulated by biologically localized embedding) nanosensors have been developed for dissolved oxygen using organically modified silicate (ormosil) nanoparticles as a matrix. The ormosil nanoparticles are prepared through a sol-gel-based process, which includes the formation of core particles with phenyltrimethoxysilane as a precursor followed by the formation of a coating layer with methyltrimethoxysilane as a precursor.

The highly permeable structure and the hydrophobic nature of the ormosil nanoparticles, as well as their small size, result in an excellent overall quenching response to dissolved oxygen and a linear response over the whole range, from 0 - 100per cent oxygen-saturated water.

This Pebble sensor has a higher sensitivity and a broader linearity as well as longer excitation and emission wavelengths, resulting in reduced background noise for cellular measurement. The Pebble sensors are excellent in terms of their reversibility and stability to leaching and long-term storage. A real-time monitoring of changes in the dissolved oxygen due to cell respiration in a closed chamber was made by gene gun delivered PEBBLE. This sensor is now being applied for simultaneous intracellular measurements of oxygen and glucose.

Application of Nanotechnology to P450 Enzymes

Cytochromes P450 are highly relevant to the bio-analytical area. They form a large family of enzymes present in all tissues essential to the metabolism of most drugs in use today, playing a vital role in the drug development and discovery process.

They act as catalysts for the insertion of one of the two atoms of an oxygen molecule into a variety of substrates (R) with quite broad regioselectivity, leading to concomitant reduction of the other oxygen atom to water as shown in the equation below.

$$RH + O_2 + 2\vec{e} + 2H^+ \rightarrow ROH + H_2O$$

Several methods have been reported in the literature for the screening of substrate turnover by P450s in a high throughput format [80-83]. However, they all fall short of being limited to testing the activity of P450 enzymes through the detection of the conversion of a specific marker substrate, but Tsotsou et al 2002 have been able to develop a method called the alkali method, which can detect the turnover of any NAD(P)H or $NAD(P)^+$ dependent enzyme.

The progress on these research fronts and their combinations provide a powerful platform for future applications of these enzymes, with particular reference to protein array technology.

APPLICATION OF NANOTECHNOLOGY TO TISSUE ENGINEERING

Tissue engineering is based on the creation of new tissues in vitro followed by surgical placement in the body or the stimulation of normal repair in situ using bioartificial constructs or implants of living cells introduced in or near the area of damage. Though it is mainly concerned with using human material, either from the patient themselves (autologous) or from other human sources (allogeneic), material from other mammalian sources have also been applied in humans (xenogeneic).

The involvement of microelectronics or nanotechnology in creating a truly bioartificial tissue or organ that can take the place of one that is terminally diseased, such as an eye, ear, heart, or joint has been envisaged. Implantable prosthetic devices and nanoscaffolds for use in the growing of artificial organs are goals of nanotechnology researchers. Nanoengineering of hydroxyapatite for bone replacement is reasonably advanced.

In the future, we could imagine a world where medical nanodevices are routinely implanted or even injected into the bloodstream to monitor wellness and to automatically participate in the repair of systems that deviate from established norms. These nanobots could be personalized by tailoring them to patient genotype and phenotype to optimize intervention at the earliest stage in the course of disease expression.

GROWTH OF NEW ORGANS

Nanoscale building of cells can be accomplished by their programmed replication. The signals are transmitted back and forth with the instruction for the desired size and shape form the construction site. When complete instructions are finished, the organs can be grown according to the prerequisite specifications.

These organs could have the necessary DNA encoded to be compatible with the required human body immunological status. This can enhance integration of artificial structures with living tissues, presenting a more appropriate interface to biological systems.

With the advantage in absence of immune reaction unlike today's donor organ transplantation. In the years to come this can accomplish a Quantum leap in the management of organ failure disorders.

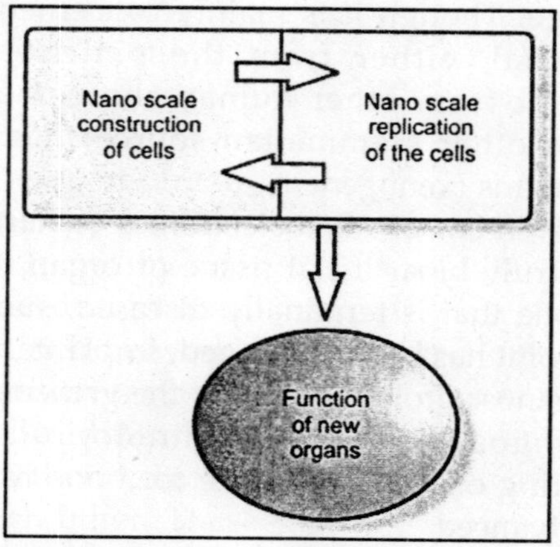

Fig. Graphical Representation of the Nanoscale Construction and Growth of new Organs

MOLECULAR IMAGING

New imaging approaches using genetically encoded fluorescent and bioluminescent reporters (i. e., illuminated or glowing identification tags) are offering revealing insights to the living body as never observed before. Information provided by these reporters can be used to enhance our understanding of human biology and the development of therapeutic approaches for many diseases, including cancer, infection, neurodegenerative and cardiovascular disease.

In addition to progresses so far made with molecular agents, industry leaders are also showcasing rapidly evolving imaging technologies that allow scientists to view organisms at the molecular level.

Table: Latest Products in Molecular Imaging and Associated Producing Companies

Product Name	Company(ies)
SPECT/CT hybrid imaging systems	Philips Medical Systems/ Siemens Medical Solutions
GFAP-luc (glial fibrillary acid protein)	Xenon

Ultrasound bubbles	Schering AG
NeuroSpec (radiodiagnostic agent)	Tyco Healthcare/ Mallinckrodt Inc.
explore Locus Ultra (Volumetric CT system)	GE Medical system
Definity or Sonolysis (nanosurgery)	ImaRx

- SPECT/CT hybrid systems capture both functional information on molecular and cellular processes (growth and activity) and anatomical detail (size and shape) of a targeted molecular structure more quickly, efficiently and clearly than standard imaging devices.
  The images obtained from these systems can assist with the rapid identification of tumours, analysis of appropriate treatment, delivery of targeted therapy to precisely destroy target cells, and follow up to assess treatment effectiveness.
- Xenon presented its newer light producing transgenic animal models (GFAP-luc) during the Society for Molecular Imaging's 3rd Annual Meeting. This model may prove to be an important model for tracking damage and repair in chronic neurological conditions such as post-ischemic stroke or Parkinson's disease.
- An ultrasound contrast agent is made of tiny "microbubbles3 that scatter light and allow the clinician to see which part of the heart muscle is poorly functioning.
  The sensitivity and flexibility of ultrasound makes it the most sensitive method of imaging microbubbles because it deliberately disrupts the pattern and produces a very strong and highly characteristic transient effect. For example,
- Definity otherwise known as Sonolysis are gas-filled microbubbles for novel therapeutic applications. For dissolving vascular thrombosis, microbubbles are

administered intravenously to a patient or injected locally into a specific vascular structure such as a vascular graft.

Ultrasound is applied externally (or can be applied internally via catheter) over the area of the blood clot to provide localized, targeted action. As the microbubbles perfuse the clot, they act as micromechanical devices where ultrasound pulses the bubbles and blows up the bubbles in the ultrasound field, leading to blood clot dissolution.

Sonolysis nanosurgery is locally targeted nanoinvasive therapy for treatment of vascular thrombosis. Compared with alternative therapies for treating thrombosis, sonolysis affords the potential merits of being less invasive than mechanical thrombectomy and faster than conventional drug therapy with less risk of bleeding.

- NeutroSpec is a radiodiagnostic agent which labels white blood cells and myeloid precursors without the need for removal and re-injection of blood into patients. This new product is for patients with equivocal signs of appendicitis who are five-years-old and up.

  NeutroSpec also facilitates the visualization of images generated via gamma camera allowing physicians to quickly and easily locate the sites of infection thereby eliminating time delays and/or risks normally affiliated with alternative white blood cell labelling processes.
- eXplore Locus Ultra is a first-class volumetric CT system capable of quantitating physiological measurements and elaborate anatomy of tissues, tumours and organ perfusion. The Locus Ultra also performs image acquisition at the rate of a sub-second, enabling dynamic imaging.

The multidisciplinary field of nanotechnology's application for discovering new molecules and manipulating those available naturally could be dazzling in its potential to improve health care. The spin-offs of nanobiotechnology could be utilised across all the countries of the world.

In the future, we could imagine a world where medical nanodevices are routinely implanted or even injected into the bloodstream to monitor health and to automatically participate in the repair of systems that deviate from the normal pattern. The continued advancement in the field of biomedical nanotechnology is the establishment and collaboration of research groups in complementary fields.

Such collaborations have to be maintained not only on specialty field level, but internationally as well. The successful development and implementation of international collaborations fosters a global perspective on research and brings together the benefits to mankind in general.

However, nanotechnology in medicine faces enormous technical hurdles in that long delays and numerous failures are inevitable. Likewise, it should not be taken for granted the dangers and negative consequences of nanobiotechnology when applied in warfare, in the hands of terrorists and disasters associated with its application in energy generation when and wherever it strikes or the risks associated with nanoparticles in blood circulation.

It should be appreciated that nanotechnology is not in itself a single emerging scientific discipline but rather a meeting point of traditional sciences like chemistry, physics, biology and materials science to bring together the required collective knowledge and expertise required for the development of these novel technologies.

Chapter 7

Nanofabrication

NANOTECHNOLOGY WORK FOR INDUSTRY

The NanoFabrication Facility (NanoFab) is one of the first Italian laboratories devoted to the transfer of nanotechnologies to industrial production. It was inspired by the Penn State Nanofabrication facility (Pennsylvania, USA) and has been created with an investment of over €25. 5m co-financed by the European Regional Development Fund (ERDF), the Italian Government, Veneto region and VEGA (the science and technology park of Venice). Veneto's traditional and low R&D intensive industrial sectors, as well as its heavy industry (notably chemicals) experienced a crisis starting from the early 1990s.

This was a consequence, on one hand, of the restructuring of large, publicly owned, companies and, on the other hand, the advent of globally integrated markets, which eroded the competitive advantage of the industrial base. The mission of NanoFab is to transfer technological knowledge and results of industrial research to firms, especially local companies, which may benefit from the application of nanotechnologies. The project's expected impact is to contribute to foster structural change in the territory's productive fabric and allow the local economy to move towards knowledge-based production.

The NanoFab is not an isolated initiative: other actions are underway to strengthen the regional nanotechnology vocation. Three new research centres are now "under construction": LaNN (focusing on developing nano-devices and lab-on chip); Ecsin (analysing the impact of nanotechnologies on the environment, health and society); Nanocomp (studying thermosetting polymers). Their aims are to complement competences already present in the region and are strongly supported by universities and industry.

NanoFab carries out, promotes and coordinates, research and technological activities through its own structures and in collaboration with universities and other public and private bodies. It provides a diverse range of high level technological services in response to specific and complex demands: companies may commission R&D projects, draw upon the scientific and technological know-how of qualified researchers and use the equipment of the centre.

The partnership that provides the foundation for NanoFab is a consortium composed of VEGA, Veneto Nanotech (a "technological district"') and the CIVEN association of Veneto universities working in the field of nanotechnology. There is a clear division of roles among these partners. Veneto Nanotech promotes NanoFab activities and research outcomes of the "nanotechnology district". CIVEN provides training courses for managers and staff, designs projects and carries out applied research relevant to industrial needs.

VEGA provided the nanofabrication facility with initial resources and infrastructures. Overall, the partnership is rather agile and the main interest for each partner participating was creating a facility which could provide high quality services related to nanotechnology in a time frame compatible with the tight requirements of private firms which want to be first movers in their markets.

The sectors of activity of NanoFab were chosen on the basis of a number of preliminary studies partly carried-out and partly commissioned by the universities. However, some important fields of application of nanotechnologies emerged only later, after the beginning of the fabrication facility's activity, and were directly suggested by the firms. For example, application of nanotechnology to textile and leather industry is now one of the most relevant aspects of NanoFab's work in Veneto.

NanoFab is a very young project, yet it has already produced outstanding results. During its first year of life (2006), the facility participated in approximately 40 projects designed by regional companies.

NanoFab is involved in 36 projects funded by the Ministry of University and Research with a consultancy role. Hence, it does not receive public funding directly rather companies that benefit from public aid schemes purchase its specialised services.

In addition, NanoFab also participated in three projects submitted to the 7th EU Research Framework Programme in 2007. Alongside the consultancy role in industrial research projects, NanoFab also receives orders from companies for undertaking feasibility studies. About 300 enterprises are negotiating with or have been in contact with NanoFab. In summary, the elements of planning, governance, management and implementation that contributed to the success of the project are:

- A winning organisational structure; the partnership between VEGA, Veneto Nanotech, CIVEN and NanoFab allowed a flexible and efficient division of competences and tasks; the capacity of NanoFab to respond to business demand quickly (thanks to its own resources and infrastructures) is the most important aspect;
- A private managerial approach; the facility is managed as a private company both regarding selection of personnel and fund raising;
- An innovative theme; nanotechnologies are pervasive and have a strong potential for being applied in traditional sectors.

  Factors, which represent valuable lessons learned for the future, are the following:
- The importance of keeping political and institutional interest alive during all the stages of project implementation;
- A focus on a declining industrial area devastated by a crisis made it easier to gather the financial support

needed to carry out high capital intensive facility and to make the most of existing infrastructures;

- An effective partnership with local universities that collaborated with the initiative, and companies which suggested new fields of application of nanotechnologies during the project implementation.

PROJECT DESCRIPTION

NanoFab is a non-profit organisation managed by a consortium that includes the VEGA Science and Technology Park of Venice1, the Interuniversity Association for Nanotechnologies CIVEN (Coordinamento Interuniversitario Veneto per le Nanotecnologie per la gestione della NanoFabrication Facility)2 and Veneto Nanotech3, the first Italian technological district focused on materials.

The NanoFabrication facility is one of the first laboratories devoted to the transfer of nanotechnologies to industrial production in Italy. It occupies a surface of 2,700m2 (total office/lab space) and has been created thanks to a total investment of €25. 5m (of which €7m were directly employed for infrastructure and equipment, the rest funded mainly research projects and training) co-financed by the Italian Government, the ERDF and Veneto Region.

The overall objective of NanoFab is the transfer of technological knowledge and industrial research results to firms, especially local companies active in national and international markets of nanotechnologies. NanoFab, on the crossroad between academia and business, aims to promote a meeting of minds between the research system and industry.

NanoFab carries out, promotes and coordinates, also within EU programmes, research and technological activities through its own structures and in collaboration with universities and other public and private bodies. It provides a diverse range of high-level technological services in response to specific and complex demands. Companies may commission R&D projects, draw upon the scientific and technological know-how of qualified researchers and use the equipment of the centre.

In NanoFab, internal researchers, coordinated by a scientific director, university researchers and companies' qualified technicians can produce materials to the highest standards and performance, and design novel production lines through high-tech processes. NanoLab has three types of multidisciplinary labs (deposition and preparation, characterisation, clean rooms) equipped with advanced tools.

Nanotechnologies developed in the laboratories may find industrial application in:

- Mechanics;
- Industrial chemistry (plastics and pigments);
- Textiles;
- Leather working;
- Sensors;
- Pharmaceutics;
- natural stone (coatings/coverings).

NanoFab, thanks to the competences of CIVEN, which is an organisation oriented to training and research in the nanotech field, organises workshops for companies. These include training courses for managers and staff in order to show potential applications of nanotechnologies and update those people on the state of the art and on future prospects.

NanoFab staff provide technology transfer services and as part of these, carry out feasibility studies which concern specific critical issues that firms encountered with their products and processes. Feasibility studies aim at finding out whether the introduction of nanotechnologies can be useful and feasible.

As part of its policy to support technological districts, the Ministry of University and Research, the Ministry of Finance and the regional government signed a framework agreement in 2004. With this agreement, the national government committed €26m5 to fund industrial research and pre-competitive development projects, as well as training in the field of nanotechnology. NanoFab does not access these funds directly, but indirectly by providing consultancy services to firms.

The creation of high capital-intensive facility (like

NanoFab), requires a massive investment that could not be completely covered by regional and national funds. Hence, ERDF contributions were critical and allowed acquiring equipment quickly. Without them, project would have required a much longer time and would include the risk of missing the opportunities in technological sector characterised by high and rapid growth.

POLITICAL AND STRATEGIC CONTEXT

Veneto, in the north-east of Italy, is characterised by a strong presence of industrial districts specialised in traditional sectors. Industrial districts experienced their "golden age" in the 1970s and 1980s. However, this model of economic development based on local production systems of small firms co-operating closely together started undergoing a crisis in the 1990s, parallel to the advent of globalisation.

Moreover, a regional specialisation in traditional industries means that the intensity of R&D investments is low. This is coupled with low levels of product and process innovation, limited importance of high-tech exports and a negative net birth rate of high tech firms. The region needs to find new specialisations and re-engineer its industrial fabric, especially with respect to those areas that are in deep crisis.

The Venice area previously hosted chemical plants owned by ENI6, the national petrol and gas company, these zones also needed to be reclaimed and converted. The entire Venetian nanotech pole has benefited from the availability of spaces and infrastructures that were previously used by the chemical industry. NanoFab also benefited from the backing of ENI and public institutions.

In short, the nanofabrication facility provides two types of advantages to the region:

- It allows the economy to exploit disused industrial structures and helps to move the regional economy towards new science-based specialisations;
- It contributes to addressing deficiencies of the local innovation system by creating a link between public research and industry.

IMPLEMENTATION

Project Design and Planning

The initiative for the project of creating nanofabrication facility in Veneto came from the regional authority, local universities and VEGA. The latter is a science and technology park, linking universities, centres of excellence and the manufacturing sector in order to sharpen the competitive edge of companies in national and international markets. VEGA, which was created in 1993, has been of fundamental importance for the creation of NanoFab in 2006. It provided resources and facilities for the project start-up.

The VEGA scientific advisor and science officer at the Italian Embassy in Washington had an intuition and suggested to the local institutions that it would have been worthwhile to create a nanotechnology fabrication facility. He was inspired during several study trips to the United States of America and managed to convince the institutions in Veneto to start the implementation process that eventually led to NanoFab.

The model for the Veneto project was the Penn State Nanofabrication Facility: an open access national facility containing highly valuable, state-of-the-art equipment and supported by 14 technical staff (holders of PhDs with cross-disciplinary expertise in the areas of spectroscopy, biology, chemistry, physics, optics, electrical engineering). It focuses on materials and chemical and molecular scale technologies with unique strengths that include surface chemistry, self-assembly and fabrication and processing of novel, non-traditional materials.

Operating for over 12 years, this facility has an established record of proactively identifying and engaging non-traditional users of nanotechnology and of bringing nanoscience and engineering to new disciplines. Experienced technical staff supports academic and industrial users through comprehensive education and training on instrument operation, process development and integration, and the application of micro and nanofabrication techniques to device, system and process development.

The decision to create NanoFab was also influenced by a McKinsey study commissioned in 2000 by the university. This study pointed out the presence of high-level competences (in Veneto), which could be capitalised in order to create a nanotech hub. The study also emphasised the potential direct industrial impact as well as the spillovers that were likely to be generated by a project of this kind.

The project concerns both incremental and breakthrough innovation which is notoriously a very risky activity. The important market failures characterising the activities of carrying out frontier research and breakthrough innovation represent important justification to public intervention and strengthen the rationale of research policy. The "three-headed" managing consortium of NanoFab (which includes the VEGA park, Veneto Nanotech district and CIVEN) is, in its own right, a structure that minimises the risks connected with the provision of highly innovative services to enterprises.

Indeed, the consortium is independent from the local large universities both in terms of personnel and infrastructures. This allows for prompt responses to clients' requirements and hence compatibility with the speed of technological innovation.

On the contrary, the average response of the public sector university would be much slower, especially considering administrative procedures, which may delay the conclusion of agreements and completion of contracts. Moreover, CIVEN, which conducts research whose outcomes are then transferred to the market by NanoFab, represents all Veneto's universities and this reduces transactions costs and makes things simpler.

NanoFab can also count on its own researchers whose main preoccupation is not publishing, but rather concluding agreements with private firms. All of these aspects, carefully assessed in the project design phase, aim at minimising the risk of the project being unsustainable and incapable of generating revenues.

Management, Monitoring and Evaluation System

The shareholders have high degree of independence in

managing the project. The regional authority initiated the process and participates in its evaluation without playing a major part in the technical decisions on research projects and in the selection of project manager and staff. The path leading to the creation of NanoFab comprised the following steps:

- 12 April 2002 - Regional resolution for the creation of an industrial research lab;
- 20 December 2002 - VEGA receives €4. 6m from the EU and co-finance creation of NanoFab with an additional investment of €2. 4m;
- 17 March 2003 - Constitution of CIVEN, association between the University of Padova, University of Venice - Ca' Foscari, University of Verona;
- 31 July 2003 - Constitution of Veneto Nanotech SCpA;
- 21 December 2004 - Constitution of NanoFab Soarl, management organisation participated in by VEGA and CIVEN;
- 7 October 2006 - Official inauguration of NanoFab.

The evaluation of results has been taken into consideration since the project's inception. The scientific directors of CIVEN assessed technical aspects and scientific outcomes. A technical regional committee also played significant role as it carried out its own evaluation alongside an assessment by CIVEN. The staff of NanoFab, including scientific directors, is selected using a private sector approach. NanoFab usually advertises posts, including on their website, and then shortlist candidates.

Governance: Partnership and Leadership

The partnership, which lies at the roots of NanoFab, is a consortium composed of the VEGA S&T Park, Veneto Nanotech and CIVEN university association. There is a clear separation of roles among these partners and this is described in the following paragraphs. Veneto Nanotech is a technological district created by the Ministry of University and Research in 2003.

It is a public-private company, which liaises with public institutions such as the regional administration and the ministry. Its main function is promotion of NanoFab activities and research outcomes of the "nanotechnology district". Veneto Nanotech also arranges conferences and recently it began to organise an annual 'nanochallenge' with an award of €100,000 to a firm which plans to invest in the region in order to carry out particularly innovative projects on nanotechnology.

CIVEN is the association of Veneto universities, which is financially supported by the regional authority. Civen monitors the latest developments in the area of nanotechnology, design projects and carries out applied research relevant to industrial needs. CIVEN also organises and manages a Master of Science in nanotechnology and trains young researchers by means of PhD scholarships. This organisation does not rely on university's human resources and equipment, but has its own staff and infrastructures.

The VEGA Science Park is a non-profit, limited consortium, which was established in 1993 and is comprised of 34 partners. Among them are local administrations, universities of Venice, the most important local private firms, two banks and a number of SMEs. The idea of NanoFab, as it was stressed previously, came from VEGA, which also provided the nanofabrication facility with initial resources and infrastructures.

Finally, the partnership includes the regional administration, which is in charge of evaluating the results of NanoFab through its technical committee. In the middle of 2007, the Veneto Chamber of Commerce also joined this partnership. NanoFab, supported by the partners, manages the nanotech labs, works directly with firms and transfers the CIVEN research outcomes to enterprises.

Overall, the partnership is quite agile and the main interest for each partner to participate was creating a facility that could provide high quality services related to nanotechnology in a time frame compatible with the "competitive rush" of private firms. Universities, often characterised by very slow reaction times due to administrative constraints, participate in the project only indirectly, through CIVEN, which is entirely devoted and focused on nanotechnology-applied research.

Political support accompanied the initiative from the very beginning of the project. In this respect, the liaison and promotion effort made by Veneto Nanotech has been crucial in fostering the interest of local and national institutions with respect to the activities carried out by CIVEN and NanoFab.

Innovative Elements and Novel Approaches to Implementation

The partners had collaborated before the creation of NanoFab, but not to the same extent. To create an effective nanofabrication facility, which could respond in a timely manner to the businesses interested in developing nanotechnology solutions, an innovative approach both in terms of organisational arrangements and financing was necessary.

NanoFab's establishment was co-funded by the ERDF, under the Veneto Single Programming Document, but also supported by the regional administration with its own budget and by the Ministry for University and Research. It is worth emphasising the novelty of the funding approach that is not direct (at least after the initial period when the equipment was acquired). NanoFab provides a range of services to enterprises that participate in publicly financed research projects. Therefore, NanoFab has access to public resources only through contract research or consulting orders.

The Ministry of University and Research significantly supports the undertaking of industrial research projects to be carried out by regional firms in the fields related to nanotechnology. NanoFab advises and provides specialised services to most of the companies participating in these projects.

Project can benefit from the promotion carried out by Veneto Nanotech, which is sponsoring the entire nanotechnology pole at regional, national and international levels. There is also an opportunity to present NanoFab activities during the dissemination days organised by CIVEN, when research projects carried out by the university association are completed. Due to the confidential nature of the work done with firms, during promotion events Nanofab can only sketch out its nanotech solutions without going into detail. NanoFab is seeking permission from its clients in order to be able to show at least their brand names during public presentations.

The need for confidentiality has important consequences on exploitation of results. Firms would like to retain most of the aspects of their collaboration with NanoFab as commercial secrets. On the other hand, NanoFab's mission is encouraging transfer and diffusion of knowledge.

This goal is ensured through special agreements which firms are required to sign with NanoFab at the beginning of their partnerships. Basically, the fabrication facility promises not to disclose specific information about customised solutions developed for individual firms, but obtains the permission to disseminate more general findings and knowledge that can produce positive externalities.

Key Implementation Obstacles and Problem-Solving Practices

The project is complex and ambitious and as a consequence its implementation has been characterised by several critical aspects. Activities related to technology transfer are very sensitive and require a great deal of confidentiality. However, NanoFab still has to sort out some issues related to intellectual property management in order to find a balance between what remains a commercial secret and what can be disclosed in order to boost knowledge spillovers in the field of nanotechnology.

The sectors of activity of NanoFab were chosen on the basis of a number of preliminary studies carried out and commissioned by universities. The McKinsey report was particularly influential in pushing towards the creation of a nanotech hub and helped identify regional sectors particularly relevant for nanotech research.

However, some important fields of application of nanotechnologies (e. g. the textile and leather industry) emerged only later, after the beginning of the fabrication facility's activity. Perhaps a more accurate initial analysis of scenarios would have identified all of the most important applications fields in Veneto. This demonstrates the ability of the project to orient its activities according to firms' needs. Reinforcing the relationship between local industry and public research was indeed one of the project's objectives.

INNOVATION RESULTS

NanoFab represents an example of indirect support to innovation and addresses deficiencies in the regional innovation system, such as weak relationship between universities and firms, as well as the structural weaknesses of the local productive fabric.

The NanoFab project contributed simultaneously to enhancing strategy and policy making capacity since, when it started, the first attempt was made to identify future scenarios for the application of nanotech solutions; reinforcing the links between different actors in the innovation system; fostering the development of innovation support services such as intellectual property rights and knowledge transfer.

NanoFab is a very young project, yet it has already produced excellent results. During its first year of existence (2006), the facility participated in approximately 40 projects designed by regional companies. Some 36 projects, of a total value of approximately €90m, have been funded by the Italian Ministry of University and Research in relation to the last national nanotechnology call for proposals in 2006.

NanoFab is involved in these projects in an advisory capacity and hence does not receive public funding directly, but is paid by companies, which benefit from aid schemes, for the provision of specialised services. More recently, NanoFab also participated in three projects submitted to the European Commission's 7th Research Framework Programme.

In addition to the advisory role in industrial research projects, NanoFab receives orders from companies to carry out feasibility studies. About 300 companies are negotiating with, or have been in contact, with NanoFab. Many important companies, active in "made in Italy" types of traditional industries, but also in automotive and mechanics are collaborating with the facility. In the Italian context, this outcome is remarkable in the area of technology transfer.

From a technical point of view, NanoFab enabled the:

- Creation of 2,700m^2 of labs, technical offices, etc.
- Acquisition of 80 scientific equipment for the production of nanomaterials, and the control of research quality;
- Creation of three "white chambers".

NanoFab has accumulated wide range of competences and know-how since the beginning of its activity.

These are transferred to firms and include the provision of solutions for:

- Nanostructured coatings;
- Nanocomposite polymers with improved thermo-mechanic properties;
- Chemical and biochemical sensoring for environmental monitoring;
- Sintered materials with improved performance/costs ratio.

Other competences are being developed, for instance:

- Innovative techniques for the treatment of leather and natural fabrics;
- Anti-reflex surfaces "moth-eye";
- Cold spray technical coatings;
- Sensoring for biological, agricultural and food productions.

In terms of results foreseen, NanoFab intends to become the main regional reference for scientific know-how, technical competences and practical solutions for innovation and development in nanotechnology. It aims to provide technology transfer services both through its own structures and through coordination with the local and international academic institutions. The final impact expected by NanoFab is to contribute fostering structural change in the territory's productive fabric and to allow switching of the local economy from traditional sectors to new knowledge-based specialisations.

It is also worth noting that NanoFab is just one but not

the only action taken to strengthen the regional nanotechnology vocation. Three new high-tech laboratories are now "under construction" in Veneto:

- LaNN – a nanotech and nano-fabrication lab will be based on the close collaboration among the technological district, Padova University and National Research Council. It will focus on developing nano-devices and lab-on chip;
- Ecsin – a lab that is now in its start-up phase and will focus on the impact of nanotechnologies on the environment, health and society;
- Nanocomp – a research centre on thermosetting polymers that aims to complement the competences already present on the territory. It is strongly supported by the Vicenza departments of Padova University and by industry.

SUSTAINABILITY AND TRANSFERABILITY

Sustainability

The ERDF was very important for the development of VEGA, the first brick that led to the creation of NanoFab. The project was inspired by the need to convert the industrial area of Venice and introduce a different type of specialisation. Regarding NanoFab, ERDF funding was crucial for the start-up phase.

The project has been designed, however, in a way to optimise its chances to quickly become sustainable.

The fact that it does not get direct funding, but can catalyse resources only thanks to services provided to companies carrying out industrial research, is a strong push towards self-sustainability. The facility is expected to reach break-even point by 2009.

Transferability

In theory, the project is applicable to other regions if there is political and institutional support for creating such a facility. However, in order to be able to create a nanofabrication facility, it is important that regional research system is equipped with the relevant competences. Moreover, it is necessary, in the launch phase, to make an effort to correctly identify regional strengths, in terms of sectors that would be able to directly embrace the outcomes of a facility such as NanoFab, or benefit indirectly from its spillovers.

Something that can be learned from NanoFab is the importance of being able to provide high-level services to private companies, promptly and hence compatibly with the rapid pace of innovation in highly competitive markets. The agile and flexible partnership, on which NanoFab is based, plays an important role and this should be taken into account by other regions wanting to repeat similar experience in their territories.

Obviously, policy makers wanting to replicate NanoFab, or similar experience in other territories, should carefully assess the actual and potential local research capacities. In case of nanotechnology, for example, it would be crucial to appraise the local competences and research capacities in relation with chemistry, engineering of microsystems and advanced materials.

It is also important to assess the potential market, that is exploring possible productive domains where the activities of project, such as NanoFab, can be applied. This effort was done in Veneto before the launch of NanoFab, but probably not thoroughly enough since some important fields of application of nanotechnologies, which can drive applied research, carried out or commissioned by the centre, emerged only later (e. g. textiles) and were suggested by firms.

MAIN SUCCESS FACTORS

NanoFab, together with the other initiatives that constitute the Venetian nanotechnology pole, can be considered as a very successful project.

However, given its "young age", it is not possible to assess the full results completely. The elements of planning, governance, management and implementation, which contributed to the success of the project.

Winning Organisational Arrangement

The partnership between VEGA, Veneto Nanotech, CIVEN and NanoFab allowed a flexible and efficient division of competences and tasks. The most important aspect was the capacity of NanoFab to respond to companies' demand quickly, thanks to its own resources and infrastructures.

Private Managerial Approach

The facility is managed as a private company both in terms of personnel selection and fund raising. In relation to staff selection,

NanoFab is directly in charge of recruiting and carries out this task by publishing vacancies on its website. In the selection of managers and researchers, great attention is given to prior experience of candidates in the private sector. In relation to the issue of fund raising, NanoFab accesses resources only by means of selling high level specialised services to private companies.

Innovative Theme

Nanotechnologies are pervasive and have a wide potential for being applied in traditional sectors. The collaborative research that, thanks to NanoFab, has taken place in textile and leather, plastics, mechanics and motorcycle production, is particularly significant. The following factors, which can be considered of key importance for NanoFab success to date, represent valuable lessons learned for the future:

- The importance of keeping political and institutional interest alive during all the stages of project implementation.
  This was achieved, in the case of NanoFab, with a continuous promotion effort, and also with some initiatives to raise the visibility of the region internationally, for example, the Master's Course on Nanotechnology, organised by CIVEN in English, to attract talents and reverse the brain drain.
- The focus on a declining industrial area devastated

by a crisis. Reclaiming space and infrastructures previously used by ENI and its chemical branches, contributed to revitalise the entire area without the need to occupy new land.

This is helping the region to find new specialisations and sets the bases for a novel development trajectory with possible important long-term growth and employment impacts on the territory.

- The effective partnership with local universities, which collaborated with the initiative. Universities provided research competences through CIVEN without, however, being involved in the management of the facility.

Chapter 8

Cellular Nanosensor

Nanosensors are any biological, chemical, or physical sensory points used to convey information about nanoparticles to the macroscopic world. Though humans have not yet been able to synthesize nanosensors, predictions for their use mainly include various medicinal purposes and as gateways to building other nanoproducts, such as computer chips that work at the nanoscale and nanorobots. Presently, there are several ways proposed to make nanosensors, including top-down lithography, bottom-up assembly, and molecular self-assembly.

Predicted Applications

Medicinal uses of nanosensors mainly revolve around the potential of nanosensors to accurately identify particular cells or places in the body in need.

By measuring changes in volume, concentration, displacement and velocity, gravitational, electrical, and magnetic forces, pressure, or temperature of cells in a body, nanosensors may be able to distinguish between and recognize certain cells, most notably those of cancer, at the molecular level in order to deliver medicine or monitor development to specific places in the body. In addition, they may be able to detect macroscopic variations from outside the body and communicate these changes to other nanoproducts working within the body.

One example of nanosensors involves using the fluorescence properties of cadmium selenide quantum dots as sensors to uncover tumors within the body. By injecting a body with these quantum dots, a doctor could see where a tumor or cancer cell was by finding the injected quantum dots, an easy process because of their fluorescence.

Developed nanosensor quantum dots would be specifically constructed to find only the particular cell for which the body was at risk.

A downside to the cadmium selenide dots, however, is that they are highly toxic to the body. As a result, researchers are working on developing alternate dots made out of a different, less toxic material while still retaining some of the fluorescence properties.

In particular, they have been investigating the particular benefits of zinc sulfide quantum dots which, though they are not quite as fluorescent as cadmium selenide, can be augmented with other metals including manganese and various lanthanide elements.

In addition, these newer quantum dots become more fluorescent when they bond to their target cells. Potential predicted functions may also include sensors used to detect specific DNA in order to recognize explicit genetic defects, especially for individuals at high-risk and implanted sensors that can automatically detect glucose levels for diabetic subjects more simply than current detectors.

DNA can also serve as sacrificial layer for manufacturing CMOS IC, integrating a nanodevice with sensing capabilities. Therefore, using proteomic patterns and new hybrid materials, nanobiosensors can also be used to enable components configured into a hybrid semiconductor substrate as part of the circuit assembly. The development and miniaturization of nanobiosensors should provide interesting new opportunities.

Other projected products most commonly involve using nanosensors to build smaller integrated circuits, as well as incorporating them into various other commodities made using other forms of nanotechnology for use in a variety of situations including transportation, communication, improvements in structural integrity,

And robotics. Nanosensors may also eventually be valuable as more accurate monitors of material states for use in systems where size and weight are constrained, such as in satellites and other aeronautic machines.

Existing Nanosensors

Currently, the most common mass-produced functioning nanosensors exist in the biological world as natural receptors of outside stimulation. For instance, sense of smell, especially in animals in which it is particularly strong, such as dogs, functions using receptors that sense nanosized molecules. Certain plants, too, use nanosensors to detect sunlight; various fish use nanosensors to detect minuscule vibrations in the surrounding water; and many insects detect sex pheromones using nanosensors.

Certain electromagnetic sensors have also been in use in photoelectric systems. These work because the specific sensors called, aptly, photosensors are easily influenced by light of various wavelengths. The electromagnetic source transfers energy to the photosensors and energizes them into an excited state which causes them to release an electron into a semiconductor. At that point, it is relatively easy to detect the electricity coming from the sensors, and thus easy to know if the sensors are receiving light.

Though more advanced uses of photosensors incorporating other forms of nanotechnology have yet to be implemented into consumer society, most film cameras have used photosensors at the nano size for years. Traditional film uses a layer of silver ions that become excited by solar energy and clump into groups, as small as four atoms apiece in some cases, that scatter light and appear dark on the frame. Various other types of film can be made using a similar process to detect other specific wavelengths of light, including x-rays, infrared, and ultraviolet.

One of the first working examples of a synthetic nanosensor was built by researchers at the Georgia Institute of Technology in 1999. It involved attaching a single particle onto the end of a carbon nanotube and measuring the vibrational frequency of the nanotube both with and without the particle.

The discrepancy between the two frequencies allowed the researchers to measure the mass of the attached particle.

Chemical sensors, too, have been built using nanotubes to detect various properties of gaseous molecules. Carbon nanotubes have been used to sense ionization of gaseous molecules while nanotubes made out of titanium have been employed to detect atmospheric concentrations of hydrogen at the molecular level.

Many of these involve a system by which nanosensors are built to have a specific pocket for another molecule. When that particular molecule, and only that specific molecule, fits into the nanosensor, and light is shone upon the nanosensor, it will reflect different wavelengths of light and, thus, be a different colour.

Production Methods

There are currently several hypothesized ways to produce nanosensors. Top-down lithography is the manner in which most integrated circuits are now made. It involves starting out with a larger block of some material and carving out the desired form. These carved out devices, notably put to use in specific microelectromechanical systems used as microsensors, generally only reach the micro size, but the most recent of these have begun to incorporate nanosized components.

Another way to produce nanosensors is through the bottom-up method, which involves assembling the sensors out of even more minuscule components, most likely individual atoms or molecules.

This would involve moving atoms of a particular substance one by one into particular positions which, though it has been achieved in laboratory tests using tools such as atomic force microscopes, is still a significant difficulty, especially to do en masse, both for logistic reasons as well as economic ones. Most likely, this process would be used mainly for building starter molecules for self-assembling sensors.

The third way, which promises far faster results, involves self-assembly, or "growing" particular nanostructures to be used as sensors. This most often entails one of two types of assembly. The first involves using a piece of some previously created or naturally formed nanostructure and immersing it in free atoms of its own kind.

After a given period, the structure, having an irregular surface that would make it prone to attracting more molecules as a continuation of its current pattern, would capture some of the free atoms and continue to form more of itself to make larger components of nanosensors.

The second type of self-assembly starts with an already complete set of components that would automatically assemble themselves into a finished product. Though this has been so far successful only in assembling computer chips at the micro size, researchers hope to eventually be able to do it at the nanometer size for multiple products, including nanosensors.

Accurately being able to reproduce this effect for a desired sensor in a laboratory would imply that scientists could manufacture nanosensors much more quickly and potentially far more cheaply by letting numerous molecules assemble themselves with little or no outside influence, rather than having to manually assemble each sensor.

WHAT ARE NANOROBOTS

Nanorobotics is emerging as a demanding field dealing with miniscule things at molecular level. Nanorobots are quintessential nanoelectromechanical systems designed to perform a specific task with precision at nanoscale dimensions. Its advantage over conventional medicine lies on its size. Particle size has effect on serum lifetime and pattern of deposition. This allows drugs of nanosize to be used in lower concentration and has an earlier onset of therapeutic action.

It also provides materials for controlled drug delivery by directing carriers to a specific location. The typical medical nanodevice will probably be a micron-scale robot assembled from nanoscale parts. These nanorobots can work together in response to environment stimuli and programmed principles to produce macro scale results.

Elements of Nanorobots

Carbon will likely be the principal element comprising the bulk of a medical nanorobot, probably in the form of diamond or diamondoid/fullerene nanocomposites. Many other light elements such as hydrogen, sulfur, oxygen, nitrogen, fluorine, silicon, etc. will be used for special purposes in nanoscale gears and other components.

The chemical inertness of diamond is proved by several experimental studies. One such experiment conducted on mouse peritoneal macrophages cultured on DLC showed no significant excess release of lactate dehydrogenase or of the lysosomal enzyme beta N-acetyl-D-glucosaminidase an enzyme known to be released from macrophages during inflammation. Morphological examination revealed no physical damage to either fibroblasts or macrophages, and human osteoblast like cells confirming the biochemical indication that there was no toxicity and that no inflammatory reaction was elicited in vitro.

The smoother and more flawless the diamond surface, the lesser is the leukocyte activity and fibrinogen adsorption. Interestingly, on the rougher "polished" surface, a small number of spread and fused macrophages were present, indicating that some activation had occurred. The exterior surface with near-nanometer smoothness results in very low bioactivity. Due to the extremely high surface energy of the passivated diamond surface and the strong hydrophobicity of the diamond surface, the diamond exterior is almost completely chemically inert.

Constituents and Design of Nanorobots

Nanorobots will possess full panoply of autonomous subsystems whose design is derived from biological models. Drexler evidently was the first to point out, in 1981, that complex devices resemble biological models in their structural components. The various components in the nanorobot design may include onboard sensors, motors, manipulators, power supplies, and molecular computers.

Perhaps the best-known biological example of such

molecular machinery is the ribosome the only freely programmable nanoscale assembler already in existence. The mechanism by which protein binds to the specific receptor site might be copied to construct the molecular robotic arm.

The manipulator arm can also be driven by a detailed sequence of control signals, just as the ribosome needs mRNA to guide its actions.

These control signals are provided by external acoustic, electrical, or chemical signals that are received by the robot arm via an onboard sensor using a simple "broadcast architecture a technique which can also be used to import power. the biological cell may be regarded as an example of a broadcast architecture in which the nucleus of the cell send signals in the form of mRNA to ribosomes in order to manufacture cellular proteins.

Assemblers are molecular machine systems that could be described as systems capable of performing molecular manufacturing at the atomic scale which require control signals provided by an onboard nanocomputer This programmable nanocomputer must be able to accept stored instructions which are sequentially executed to direct the manipulator arm to place the correct moiety or nanopart in the desired position and orientation, thus giving precise control over the timing and locations of chemical reactions or assembly operations.

Construction of Nanorobots

There are two main approaches to building at the nanometer scale: positional assembly and self-assembly. In positional assembly, investigators employ some devices such as the arm of a miniature robot or a microscopic set to pick up molecules one by one and assemble them manually. In contrast, self-assembly is much less painstaking, because it takes advantage of the natural tendency of certain molecules to seek one another out.

With self-assembling components, all that investigators have to do is put billions of them into a beaker and let their natural affinities join them automatically into the desired configurations. Making complex nanorobotic systems requires manufacturing techniques that can build a molecular structure via computational models of diamond mechanosynthesis. DMS is the controlled addition of carbon atoms to the growth surface of a diamond crystal lattice in a vacuum-manufacturing environment. Covalent chemical bonds are formed one by one as the result of positionally constrained mechanical forces applied at the tip of a scanning probe microscope apparatus, following a programmed sequence.

Recognition of Target Site by Nanorobots

Different molecule types are distinguished by a series of chemotactic sensors whose binding sites have a different affinity for each kind of molecule. The control system must ensure a suitable performance. It can be demonstrated with a determined number of nanorobots responding as fast as possible for a specific task based scenario. In the 3D workspace the target has surface chemicals allowing the nanorobots to detect and recognize it.

Manufacturing better sensors and actuators with nanoscale sizes makes them find the source of release of the chemical. Nanorobot Control Design (NCD) simulator was developed, which is software for nanorobots in environments with fluids dominated by Brownian motion and viscous rather than inertial forces.

First, as a point of comparison, the scientists used the nanorobots' small Brownian motions to find the target by random search. In a second method, the nanorobots monitor for chemical concentration significantly above the background level. After detecting the signal, a nanorobot estimates the concentration gradient and moves toward higher concentrations until it reaches the target.

In the third approach, nanorobots at the target release another chemical, which others use as an additional guiding signal to the target. With these signal concentrations, only nanorobots passing within a few microns of the target are likely to detect the signal.

Thus, we can improve the response by having the nanorobots maintain positions near the vessel wall instead of floating throughout the volume flow in the vessel from monitoring the concentration of a signal from others; a nanorobot can estimate the number of nanorobots at the target. So, the nanorobot uses this information to determine when enough nanorobots are at the target, thereby terminating any additional "attractant" signal a nanorobot may be releasing. It is found that the nanorobots stop attracting others once enough nanorobots have responded.

The amount is considered enough when the target region is densely covered by nanorobots. Thus these tiny machines work at the target site accurately and precisely to that extent only to which it is designed to do.

Nanorobots Evading the Immune System

Every medical nanorobot placed inside the human body will encounter phagocytic cells many times during its mission. Thus all Nanorobots, which are of a size capable of ingestion by phagocytic cells, must incorporate physical mechanisms and operational protocols for avoiding and escaping from phagocytes. The initial strategy for medical nanorobots is first to avoid phagocytic contact or recognition.

To avoid being attacked by the host's immune system, the best choice is to have an exterior coating of passive diamond. The smoother and flawless the coating, the lesser is the reaction from the body's immune system. And if this fails then to avoid it's binding to the phagocyte surface that leads to phagocytic activation. If trapped, the medical nanorobot can induce exocytosis of the phagosomal vacuole in which it is lodged or inhibit both phagolysosomal fusion and phagosome metabolism.

In rare circumstances, it may be necessary to kill the phagocyte or to blockade the entire phagocytic system. The most direct approach for a fully functional medical nanorobot is to employ its motility mechanisms to locomote out of, or away from, the phagocytic cell that is attempting to engulf it. This may involve reverse cytopenetration, which must be done cautiously (e. g., the rapid exit of nonenveloped viruses from cells can be cytotoxic).

It is possible that frustrated phagocytosis may induce a localized compensatory granulomatous reaction. Medical nanorobots therefore may also need to employ simple but active defensive strategies to forestall granuloma formation. Metabolizing local glucose and oxygen for energy can do the powering of the nanorobots.

In a clinical environment, another option would be externally supplied acoustic energy. When the task of the nanorobots is completed, they can be retrieved by allowing them to exfuse themselves via the usual human excretory channels or can also be removed by active scavenger systems.

Nanorobots in Cancer Detection and Treatment

The development of nanorobots may provide remarkable advances for diagnosis and treatment of cancer. Nanorobots could be a very helpful and hopeful for the therapy of patients, since current treatments like radiation therapy and chemotherapy often end up destroying more healthy cells than cancerous ones.

From this point of view, it provides a non-depressed therapy for cancer patients. The Nanorobots will be able to distinguish between different cell types that is the malignant and the normal cells by checking their surface antigens (they are different for each type of cell).

This is accomplished by the use of chemotactic sensors keyed to the specific antigens on the target cells. Another approach uses the innovative methodology to achieve decentralized control for a distributed collective action in the combat of cancer. Using chemical sensors they can be programmed to detect different levels of E-cadherin and beta-catenin in primary and metastatic phases. Medical nanorobots will then destroy these cells, and only these cells. The following control methods were considered:

- Random: nanorobots moving passively with the fluid reaching the target only if they bump into it due to Brownian motion.
- Follow gradient: nanorobots monitor concentration intensity for E-cadherin signals, when detected,

measure and follow the gradient until reaching the target. If the gradient estimate subsequent to signal detection finds no additional signal in50ms, the nanorobot considers the signal to be a false positive and continues flowing with the fluid.

- Follow gradient with attractant: nanorobots arriving at the target, they release in addition a different chemical signal used by others to improve their ability to find the target.

Thus, a higher gradient of signal intensity of E-cadherin is used as chemical parameter identification in guiding nanorobots to identify malignant tissues. Integrated nanosensors can be utilized for such a task in order to find intensity of E-cadherin signals. Thus they can be employed effectively for treating cancer.

Nanorobots IN Cancer Detection and Treatment

Pharmacyte is a self-powered, computer controlled medical nanorobot system capable of digitally precise transport, timing, and targeted-delivery of pharmaceutical agents to specific cellular and intracellular destinations within the human body.

Pharmacytes escape the phagocytic process as they will not embolize small blood vessels because the minimum viable human capillary that allows passage of intact erythrocytes and white cells is 3–4 micronmeter in diameter, which is larger than the largest proposed Pharmacyte.

Pharmacytes will have many applications in nanomedicine such as initiation of apoptosis in cancer cells and direct control of cell signaling processes. Pharmacytes could also tag target cells with biochemical natural defensive or scavenging systems, a strategy called "phagocytic flagging". For example, novel recognition molecules are expressed on the surface of apoptotic cells.

In the case of T lymphocytes, one such molecule is phosphatidylserine, a lipid that is normally restricted to the inner side of the plasma membrane [1m] but, after the induction of apoptosis, appears on the outside.

Cells bearing this molecule on their surface can then be recognized and removed by phagocytic cells. Seeding the outer wall of a target cell with phosphatidylserine or other molecules with similar action could activate phagocytic behaviour by macrophages, which had mistakenly identified the target cell as apoptotic substances capable of triggering a reaction by the body Pharmacytes would be capable of carrying up to approximately 1cubicmeter of pharmaceutical payload stored in onboard tanks that are mechanically offloaded using molecular sorting pumps operated under the control of an onboard computer.

Depending on mission requirements, the payload can be discharged into the proximate extracellular fluid or delivered directly into the cytosol using a transmembrane injector mechanism.

If needed for a particular application, deployable mechanical cilia and other locomotive systems can be added to the Pharmacyte to permit transvascular and transcellular mobility, thus allowing delivery of pharmaceutical molecules to specific cellular and even intracellular addresses with negligible error.

Pharmacytes, once depleted of their payloads or having completed their mission, would be recovered from the patient by conventional excretory pathways. The nanorobots might then be recharged, reprogrammed and recycled for use in a second patient who may need a different pharmaceutical agent targeted to different tissues or cells than in the first patient.

Nanorobots in the Diagnosis and Treatment of Diabetes

Glucose carried through the blood stream is important to maintain the human metabolism working healthfully, and its correct level is a key issue in the diagnosis and treatment of diabetes. Intrinsically related to the glucose molecules, the protein hSGLT3 has an important influence in maintaining proper gastrointestinal cholinergic nerve and skeletal muscle function activities, regulating extracellular glucose concentration.

The hSGLT3 molecule can serve to define the glucose

levels for diabetes patients. The most interesting aspect of this protein is the fact that it serves as a sensor to identify glucose.

The simulated nanorobot prototype model has embedded Complementary Metal Oxide semi-conductor (CMOS) nanobioelectronics. It features a size of ~2 micronmeter, which permits it to operate freely inside the body. Whether the nanorobot is invisible or visible for the immune reactions, it has no interference for detecting glucose levels in blood stream.

Even with the immune system reaction inside the body, the nanorobot is not attacked by the white blood cells due biocompatibility. For the glucose monitoring the nanorobot uses embedded chemosensor that involves the modulation of hSGLT3 protein glucosensor activity.

Through its onboard chemical sensor, the nanorobot can thus effectively determine if the patient needs to inject insulin or take any further action, such as any medication clinically prescribed. The image of the NCD simulator workspace shows the inside view of a venule blood vessel with grid texture, red blood cells (RBCs) and nanorobots. They flow with the RBCs through the bloodstream detecting the glucose levels.

At a typical glucose concentration, the nanorobots try to keep the glucose levels ranging around 130 mg/dl as a target for the Blood Glucose Levels (BGLs). A variation of 30mg/dl was adopted as a displacement range, though this can be changed based on medical prescriptions. In the medical nanorobot architecture, the significant measured data can be then transferred automatically through the RF signals to the mobile phone carried by the patient. At any time, if the glucose achieves critical levels, the nanorobot emits an alarm through the mobile phone.

Controlling Glucose Level using Nanorobots

In the simulation, the nanorobot is programmed also to emit a signal based on specified lunch times, and to measure the glucose levels in desired intervals of time.

The nanorobot can be programmed to activate sensors and measure regularly the BGLs early in the morning, before the expected breakfast time. Levels are measured again each 2 hours after the planned lunchtime. The same procedures can be programmed for other meals through the day times.

A multiplicity of blood borne nanorobots will allow glucose monitoring not just at a single site but also in many different locations simultaneously throughout the body, thus permitting the physician to assemble a whole-body map of serum glucose concentrations.

Examination of time series data from many locations allows precise measurement of the rate of change of glucose concentration in the blood that is passing through specific organs, tissues, capillary beds, and specific vessels. This will have diagnostic utility in detecting anomalous glucose uptake rates which may assist in determining which tissues may have suffered diabetes-related damage, and to what extent.

Other onboard sensors can measure and report diagnostically relevant observations such as patient blood pressure, early signs of tissue gangrene, or changes in local metabolism that might be associated with early-stage cancer. Whole-body time series data collected during various patient activities levels could have additional diagnostic value in assessing the course and extent of disease.

This important data may help doctors and specialists to supervise and improve the patient medication and daily diet. This process using nanorobots may be more convenient and safe for making feasible an automatic system for data collection and patient monitoring.

It may also avoid eventually infections due the daily small cuts to collect blood samples, possibly loss of data, and even avoid patients in a busy week to forget doing some of their glucose sampling. These Recent developments on nanobioelectronics show how to integrate system devices and cellular phones to achieve a better control of glucose levels for patients with diabetes.

Respirocyte - An Artificial Oxygen Carrier Nanorobot

The artificial mechanical red cell, "Respirocyte" is an imaginary nanorobot, floats along in the blood stream. These atoms are mostly carbon atoms arranged as diamond in a porous lattice structure inside the spherical shell. The Respirocyte is essentially a tiny pressure tank that can be pumped full of oxygen (O_2) and carbon dioxide (CO_2) molecules. Later on, these gases can be released from the tiny tank in a controlled manner.

The gases are stored onboard at pressures up to about 1000 atmospheres. Respirocyte can be rendered completely nonflammable by constructing the device internally of sapphire, a flameproof material with chemical and mechanical properties otherwise similar to diamond.

There are also gas concentration sensors on the outside of each device. When the nanorobot passes through the lung capillaries, O_2 partial pressure is high and CO_2 partial pressure is low, so the onboard computer tells the sorting rotors to load the tanks with oxygen and to dump the CO_2. When the device later finds itself in the oxygen-starved peripheral tissues, the sensor readings are reversed.

That is, CO_2 partial pressure is relatively high and O_2 partial pressure relatively low, so the onboard computer commands the sorting rotors to release O_2 and to absorb CO_2. Respirocytes mimic the action of the natural hemoglobin-filled red blood cells. But a Respirocyte can deliver 236 times more oxygen per unit volume than a natural red cell.

This nanorobot is far more efficient than biology, mainly because its diamondoid construction permits a much higher operating pressure. So the injection of a 5 cm^3 dose of 50per cent Respirocyte aqueous suspension into the bloodstream can exactly replace the entire O_2 and CO_2 carrying capacity of the patient's entire 5,400 cm^3 of blood. Respirocyte will have pressure sensors to receive acoustic signals from the doctor, who will use an ultrasound-like transmitter device to give the Respirocyte commands to modify their behaviour while they are still inside the patient's body.

Artificial Phagocytes – Microbivores Nanorobots

A microbivore has been described, whose primary function is to destroy microbiological pathogens found in the human bloodstream, using the "digest and discharge" protocol. Nanorobotic artificial hypothetical phagocytes called "microbivores" could patrol the bloodstream, seeking out and digesting unwanted pathogens including bacteria, viruses, or fungi. Microbivores when given intravenously (I. V) would achieve complete clearance of even the most severe septicemic infections in hours or less.

This is far better than the weeks or months needed for antibiotic-assisted natural phagocytic defenses. The nanorobots do not increase the risk of sepsis or septic shock because the pathogens are completely digested into harmless simple sugars, monoresidue amino acids, mononucleotides, free fatty acids and glycerol, which are the biologically inactive effluents from the nanorobot.

A Hypothetical Mobile Cell-Repair Nanorobot

Another nanorobot, the Chromallocyte would replace entire chromosomes in individual cells thus reversing the effects of genetic disease and other accumulated damage to our genes, preventing aging.

Chromallocyte is a hypothetical mobile cell-repair nanorobot capable of limited vascular surface travel into the capillary bed of the targeted tissue or organ, followed by extravasation, histonatation, cytopenetration, and complete chromatin replacement in the nucleus of one target cell, and ending with a return to the bloodstream and subsequent extraction of the device from the body, completing the cell repair mission.

Inside a cell, a repair machine will first size up the situation by examining the cell's contents and activity, and then take action. By working along molecule-by-molecule and structure-by-structure, repair machines will be able to repair whole cells. By working along cell-by-cell and tissue-by-tissue, they (aided by larger devices, where need be) will be able to repair whole organs. By working through a person organ by organ, they will restore health. Because molecular machines will be able to build molecules and cells from scratch, they will be able to repair even cells damaged to the point of complete inactivity.

Further Applications of Nanorobots

Nanorobots could be used to maintain tissue oxygenation in the absence of respiration, repair and recondition the human vascular tree eliminating heart disease and stroke damage, perform complex nanosurgery on individual cells, and instantly staunch bleeding after traumatic injury. Monitoring nutrient concentrations in the human body is a possible application of nanorobots in medicine.

One of interesting nanorobot utilization is also to assist inflammatory cells (or white cells) in leaving blood vessels to repair injured tissues.

Nanorobots might be used as well to seek and break kidney stones. Nanorobots could also be used to process specific chemical reactions in the human body as ancillary devices for injured organs. Nanorobots equipped with nanosensors could be developed to deliver anti-HIV drugs. Another important capability of medical nanorobots will be the ability to locate stenosed blood vessels, particularly in the coronary circulation, and treat them mechanically, chemically, or pharmacologically.

To cure skin diseases, a cream containing nanorobots may be used. It could remove the right amount of dead skin, remove excess oils, add missing oils, apply the right amounts of natural moisturizing compounds, and even achieve the elusive goal of 'deep pore cleaning' by actually reaching down into pores and cleaning them out. The cream could be a smart material with smooth-on, peel-off convenience.

A mouthwash full of smart nanomachines could identify and destroy pathogenic bacteria while allowing the harmless flora of the mouth to flourish in a healthy ecosystem. Further, the devices would identify particles of food, plaque, or tartar, and lift them from teeth to be rinsed away.

Being suspended in liquid and able to swim about, devices would be able to reach surfaces beyond reach of toothbrush bristles or the fibers of floss. As short-lifetime medical nanodevices, they could be built to last only a few minutes in the body before falling apart into materials of the sort found in foods.

Medical nanodevices could augment the immune system by finding and disabling unwanted bacteria and viruses. When an invader is identified, it can be punctured, letting its contents spill out and ending its effectiveness. If the contents were known to be hazardous by themselves, then the immune machine could hold on to it long enough to dismantle it more completely.

Devices working in the bloodstream could nibble away at arteriosclerotic deposits, widening the affected blood vessels. Cell herding devices could restore artery walls and artery linings to health, by ensuring that the right cells and supporting structures are in the right places. This would prevent most heart attacks.

Nanorobots could be used in precision treatment and cell targeted delivery, in performing nanosurgery, and in treatments for hypoxemia and respiratory illness, dentistry, bacteremic infections, physical trauma, gene therapy via chromosome replacement therapy and even biological aging. It has been suggested that a fleet of nanorobots might serve as antibodies or antiviral agents in patients with compromised immune systems, or in diseases that do not respond to more conventional measures.

There are numerous other potential medical applications, including repair of damaged tissue, unblocking of arteries affected by plaques, and perhaps the construction of complete replacement body organs. Nanoscale systems can also operate much faster than their larger counterparts because displacements are smaller; this allows mechanical and electrical events to occur in less time at a given speed.

Nanotechnology as a diagnostic and treatment tool for patients with cancer and diabetes showed how actual developments in new manufacturing technologies are enabling innovative works which may help in constructing and employing nanorobots most effectively for biomedical problems.

Nanorobots applied to medicine hold a wealth of promise from eradicating disease to reversing the aging process (wrinkles, loss of bone mass and age-related conditions are all treatable at the cellular level); nanorobots are also candidates for industrial applications. The advent of molecular nanotechnology will again expand enormously the effectiveness, comfort and speed of future medical treatments while at the same time significantly reducing their risk, cost, and invasiveness.

CELLULAR BIOSCANNING

The goal of cellular bioscanning is the noninvasive and non-destructive in vivo examination of interior biological structures. One of the most common nanomedical sensor tasks is the scanning of cellular and subcellular structures.

Such tasks may include localization and examination of cytoplasmic and nuclear membranes, as well as the identification and diagnostic measurement of cellular contents including organelles and other natural molecular devices, cytoskeletal structures, biochemical composition and the kinetics of the cytoplasm.

The precision and speed of medical nanodevices is so great that they can provide a surfeit of detailed diagnostic information well beyond that which is normally needed in classical medicine for a complete analysis of somatic status.

Except in the most subtle cases a malfunction in any one component of the cellular machinery normally provokes a cascade of pathological observables in many other subsystems. Detection of any one of these cascade observables, if sufficiently unique and well-defined, may provide adequate diagnostic information to plan a proper reparative procedure.

DNA is one of the few cellular components that is regularly inspected and repaired. Most homeostatic systems adopt a more simple philosophy of periodic replacement of components regardless of functionality. Nanomedicine allows the philosophy of inspection and repair to be extended to all cellular components. The following discussion briefly describes a few of the sensory techniques that may be useful in achieving this objective.

Chapter 9

Nanomedicine

The future capabilities of nanomedicine give hope and inspiration to those of us who still have decades of life to look forward to, but some are not so fortunate. Many others who rightfully should live several decades more might find that chance cuts short their expected time.

Heart attacks and cancer can strike us down even in the prime of our lives. They do not always wait their turn and politely arrive only when expected. How can today's dying patient take advantage of a future medical technology that is as yet only described in a handful of theoretical publications? How can we preserve the physical structure of our bodies well enough to permit that future medical technology to restore our health?

Living systems can usually heal and cure their own injuries, unless those injuries are severe enough to prevent the living system from functioning. Too often, we suffer injuries that are indeed this severe. Molecular nanotechnology is feasible. As we master the ability to design molecular machines that can continue to function when the living system around them has failed, those molecular machines can restore the function of the living system.

They can support and sustain the processes of the living system until that living system can once again function on its own. Whether this is done by a temporary assist from respirocytes or by any of the myriad other techniques discussed in *Nanomedicine,* the underlying message is clear: life and health can be restored and sustained in the face of greater injury, greater damage, greater trauma, and greater dysfunction than has ever before been realized. This will usher in a new era of medicine — an era in which health and long life will be the usual state of affairs while sickness, debility and death will be the mercifully rare exceptions.

The extraordinary medical prospects ahead of us have renewed interest in a proposal made long ago: that the dying patient could be frozen, then stored at the temperature of liquid nitrogen for decades or even centuries until the necessary medical technology to restore health is developed. Called cryonics, this service is now available from several companies. Because final proof that this will work must wait until after we have developed a medical technology based on the foundation of a mature nanotechnology, the procedure is experimental.

We cannot prove today that medical technology will (or will not) be able to reverse freezing injury 100 years from now. But the patient dying today must choose whether to join the experimental group or the control group. The luxury of waiting for a definitive answer before choosing is simply not available. So the decision must be made today, on the basis of incomplete information.

We already know what happens to the control group. The outcome for the experimental group has not yet been confirmed. But given the wonderful advances that we see coming, it seems likely that we should be able to reverse freezing injury — especially when that injury is minimized by the rapid introduction through the vascular system of cryoprotectants and other chemicals to cushion the tissues against further injury.

The development of nanomedicine depends on us: what we do and how rapidly we do it. Research is not done by a faceless "them," nor is it something that happens spontaneously and without any human intervention. It is done by and supported by people. Unless we decide to support and pursue this research, it won't happen. How long it takes to develop depends on us. We are not idle bystanders watching the world go by. We are a part of it.

If we sit and wait for someone else to develop this

technology, it will happen much more slowly. If we jump in and work to make it happen, it will happen sooner. And developing a life saving medical technology within our lifetimes seems like a very good idea — certainly better than the alternative.

Sometime in the next 30 years, mankind should acquire unprecedented ability to grasp, manipulate and modify individual molecules, and this ability may have profound implications. The term "molecular nanotechnology" has been coined to mean the technology of highly versatile and inexpensive molecular fabrication, molecular manipulation, and molecular-level manufacturing.

It refers to the nanometer size range, the scale of atoms and molecules, but could be used to create macroscopic structures of precisely defined composition. The key concepts of molecular nanotechnology are that molecules can be machines and that molecular engineers can work with molecules to build desired equipment just as effectively as today's engineers now work with bulk materials.

The difference between nanotechnologists and biotechnologists is that the former do not restrict themselves to the biological limitations of the latter, and they are much more ambitious about the kinds of accomplishments that they want to achieve.

Organic chemists have long been able to synthesize complex molecular structures, including drugs, by devising ingenious reagents and by contriving reaction conditions in such a way as to minimize undesired side reactions and maximize yield. The results obtained depend on the statistics of uncontrolled molecular collisions in solution involving all possible molecular degrees of freedom. Statistically improbable reactions may be slow.

More important, solution-based reactions generally lead to unwanted and potentially toxic byproducts. This imprecision of present day chemical synthesis and manufacturing processes is one reason that drug standards are needed. In contrast, molecular nanotechnology would directly and rapidly produce the desired chemical and only the desired chemical, giving yields of 100per cent. This would be achieved by precisely positioning and bringing together the individual molecules involved in the reaction in such a way as to catalyze the reaction desired and only the reaction desired.

NATURAL MOLECULAR MACHINES

Enzymes are natural molecular machines that adsorb individual reactant molecules from the surrounding solution and, as a result of precisely orienting them with respect to each other in a protected "nanoenvironment," catalyze reactions in a highly specific manner at very high speeds and under mild reaction conditions. This simple process in biological systems ultimately allows synthesis of structures as diverse as carbon dioxide and hair.

In fact, living organisms are naturally-existing, fabulously complex systems of molecular nanotechnology. If nature can produce the biochemical capabilities of living cells by accident, molecular engineers should be able to accomplish comparable, but broader capabilities by design, guided in part by the examples provided by living systems.

Many industries already make use of enzymes to catalyze desired reactions one molecule at a time. Genetic engineers are producing pharmaceuticals by using naturally occurring enzymes to edit DNA, and the soft drink industry uses enzymes that have been modified to allow them to produce sugar at high rates near 100 degrees C without denaturation. The real promise for the future, however, lies in the development of fully artificial enzymes.

Enzymes have already been designed, synthesized, and found to function as designed. Designed enzymes that are found not to function as intended can be modified as many times as necessary until they function as desired. Thus, whether from first principles or from enlightened trial and error, industrially useful artificial enzymes should be forthcoming. More than 10^{100} average-sized synthetic enzymes are possible with the use of nature's 20 amino acids, whereas probably less than 10^{14} enzymes are presently responsible for maintaining the entire biosphere.

But there is nothing to restrict artificial enzymes to only 20 amino acids. There are many useful kinds of chemistry that are not easily promoted by natural amino acids. One particularly versatile method for transcending the biological limits of proteinaceous catalysts has already been demonstrated.

It is based on the fact that the genetic code specifies amino acids by the sequence of any one of four nucleic acid bases taken three at a time, with each three-letter base sequence known as a codon. There are 64 codons in all, but nature uses them to specify only 20 amino acids rather than 64. Recently, it has become possible to create an artificial transfer RNA that can add an unnatural amino acid to a growing polypeptide chain in response to one of the "unused" codons that exist naturally.

This could be expanded to an unlimited number of unnatural amino acids, and these unnatural amino acids could contain totally nonbiological catalytic groups or even pre-made machine parts, such as structural support struts, molecular bearings, or the like. In a fully artificial system, the 20 natural amino acids might even be entirely dispensable.

Furthermore, it is now possible to insert artificial bases into DNA and RNA, drastically augmenting the prospects for designing catalytically active RNA as well as proteins with unnatural functional groups. The potential for programming the creation of unprecedented chemical catalysts and other molecular tools useful for molecular engineering is thus virtually open-ended.

Yet a ribosomal pathway to molecular nanotechnology is far from the only viable approach. Bulk techniques are now creating useful building blocks for molecular nanotechnology, such as carbon closed-end tubes with walls that are one atom thick. Cryptands or cavitands are being created that pack the catalytic punch of enzymes into nonprotein-aceous structures far smaller than natural enzymes.

At the same time, the ability of the scanning tunneling microscope not only to image, but also to manipulate individual atoms and molecules is being combined with the biological specificity of antibodies in an attempt to make all-purpose molecular synthesizers similar in concept to industrial robots that now make automobiles. These present-day efforts are being supplemented by computational chemistry simulations of molecular bearings, molecular planetary gears, and molecular robot arms that would direct molecular factories.

Molecular Medicine

An important premise of molecular nanotechnology is that the products of this technology should ultimately be readily affordable. This premise is based on the notion that an all-purpose molecular assembler should be able to make copies of itself. Thus, once the first programmable molecular assembler is made, it will not be long before there are as many such assemblers as the market needs.

Designs for computers with molecular data storage and processing elements predict data storage capacities of 1,000 megabytes per cubic micron and data-processing capabilities of 10^{10} operations per second for similar volumes. Translated into a sugar-cube-sized personal computer, this molecular computer would store 10^{20} bytes of information and process information at a speed of 10^{20} operations per second. This is 200 billion times the information storage capacity of today's top level personal computer and well over 100 million times the processing speed of today's fastest supercomputers.

Although it is difficult to project how many of the open-ended possibilities suggested by molecular nanotechnology will be realized by the year 2020, several innovations seem worth considering.

Quality control in the pharmaceutical industry today is necessary due to the imprecision of the manufacturing, purification, and packaging processes and the inability to monitor product quality on a continuous basis. When drugs are made by programmable molecular fabricators and when molecular sensing devices are available, continuous sensing of the near-flawless production process should be possible, both for internal use and for reporting purposes.

Problems could be corrected instantly by replacing defective chemical synthesizers with backup copies and discarding the few molecules of mis-synthesized material before any has a chance to leave the factory.

Thus, the products reaching the public should conform to standards with virtually exact fidelity. It is also possible that multipurpose drug synthesizers could be on hand in most clinical chemistry and toxicology laboratories and would make on-site *de novo* synthesis an alternative to obtaining pre-synthesized standards from bodies such as the USP.

The barriers to this route could be more political than technical, but the availability of rapid local synthesis could make a life-or-death difference in some cases, such as those involving acute poisoning, for example. Alternatively, and more probably, instruments capable of identifying and quantitating drugs without calibration standards should be feasible.

Smart Drugs

Today's drug is essentially a single molecule with an often sophisticated but always limited repertoire. Tomorrow's "smart pharmaceuticals." could be essentially programmable machines with a range of "sensory," "decision-making," and "effector" capabilities. They might avoid side effects and allergic reactions by coming in generic, biocompatible housings; becoming active only upon reaching their ultimate destinations; and attaining almost complete specificity of action.

They might check for over-dosage before becoming active, thus preventing accidental or intentional poisoning. They might have not one chemical action but several, processing targeted invading organisms or malignant cells through a series of chemical reactions that guarantee the death of the target. They might work in concert with three or four "sister" agents that together produce versatility unattainable by one agent alone.

Despite the vast increase in complexity over present-day drugs, such agents can be expected to be totally "pure" and predictable in their behaviour. Safety and efficacy may be inherent in the designs of such "drugs," in which case regulatory issues could be simplified rather than complicated.

Drug Information

When every intelligent person can have essentially unlimited data storage and data-processing capability at his or her fingertips, drug information will change dramatically.

Marked advances in diagnostic agents could have far-reaching consequences. The year 2020 will occur approximately 15 years after the complete human genome has been obtained. By 2020, great advances in understanding biochemical individuality, which is so important for side effects and proper dose adjustment, will have been made.

This may make it possible for the physician to read critical aspects of a patient's phenotype from a noninvasively obtained cell sample and to inject "smart" diagnostic agents that can be recovered in a drop of saliva some time later and read out to reveal signs of previously undetected, impending disease processes. The opportunities for precise tailoring of individual treatment and for preventive medicine, with all the cost savings implied by both, would be revolutionary.

"Drug information" could come to include the complete matrix of appropriate pharmacologic responses to the newly-available individuality data. Eventually, this will become so complex and extensive that the physician will be utterly dependent on assistance provided by computer "expert systems" in making decisions.

In the long run, particularly if ingestible diagnostic agents and home readout systems become available, the need for physician participation in pharmacological therapy may vanish altogether for those patients with the right equipment available.

A major incentive for this decentralization of medical care will be ethical concerns over patient privacy and the use of genotypic information for unauthorized purposes: these are issues of "drug information" that continue to redefine the term. Relevant compendia should be available on line to any interested party, particularly those whose software allows them to interface the data with a medical "expert system."

Information about "drugs" per se will also change dramatically. Drug documentation will come to resemble today's protein structural databases because of the potentially great complexity of "pharmaceutical agents" that differ from medical devices only in that they are injectable, and/or invisible to the naked eye. Indeed, the line between pharmaceuticals and medical devices will become fuzzier and fuzzier. On the other hand, the need to document contra-indications, side effects, and complications of use should be considerably reduced.

The world of health care technology will be substantially different in 2020 from that of today. The Pharmacopeial Convention will likely be a much different organization on the occasion of its bicentennial anniversary. The need for pharmaceutical scrutiny by medical practitioners could be even more intense, however, as all aspects of pharmaceutical medicine and the pharmaceutical industry itself continue to change dramatically and at great speed.

It can be hoped that molecular nanotechnology, in addition to helping to create these dramatic changes, will also be one of the technologies that help the practitioners of 2020 stay abreast of and manage these developments.

TINY POLYMERIC DRUG CARRIERS

Currently the area of nanomedicine showing the greatest potential is the use of tiny polymeric drug carriers. Such materials can now be made in the form of spheres with diameters ranging from about 50 billionths of a meter. This can facilitate the transport of drugs to some of the smallest capillaries in the body. These nanospheres are designed to travel to specific sites within the body, release their payload of drug molecules and then degrade.

Their degradation by-products are non-toxic and will ultimately be excreted from the body. Developments in polymer science have had a great influence on drug delivery. It is now possible to synthesise a wide variety of biocompatible biodegradable polymers that will release entrained drugs at a rate determined by the chemistry and physical form of the polymer. Drug delivery research is now at a very important stage in its development.

In recent years there has been revolutionary progress made in the field of healthcare. The knowledge now exists to cure many afflictions that were once thought incurable through the use of novel therapeutics. Many of these therapeutic methods cannot currently be used or are not being used to their full potential because they cannot be delivered into the body with the required control.

An example of this is chemotherapy for the treatment of cancer. This involves saturating the body with toxic drugs, which can often result in harmful side effects such as reduced immune response to infection. By directing the drug molecules specifically to the site of a tumour a reduced quantity of drug would be required thus reducing the toxic effect on the body.

Advances in biotechnology now make it possible to control disease at the genetic level. These developments have lead to the creation of several new fields in the area of biopharmaceutical science. These new therapies are based on the interaction of nucleic acid drugs with the genetic material of specific cells within the body.

The Achilles' heel of these new biopharmaceuticals is that it is difficult to direct them to the specific site required for action. There is also another disadvantage to these therapies in that they are easily broken down by conditions within the body. Incorporation of such drugs into delivery devices may be the way forward.

So how are such nanoparticulate devices produced? The answer is a complicated one that involves the use of molecules that self-assemble themselves during a chemical process. This procedure, which is known as 'emulsion polymerisation'. This involves the use of self-assembled spheres as a molecular 'mould' into which reactive monomer diffuses and polymerises to form solid spheres.

Molecules known as surfactants perform this bizarre self-assembly by virtue of their unusual structure. The process has been likened to putting transistors, resistors and diodes into a bag, shaking it around and pulling out a fully functional radio. The mechanism that causes surfactant molecules to self-assemble is a result of their amphiphillic nature.

This means that the molecules are composed of two or more parts with each part being soluble in a different medium. Generally, a surfactant may have a hydrophilic head and a hydrophobic tail.

Once the concentration of these molecules in an aqueous solution reaches a critical value they will arrange themselves so as to minimise the interaction of the hydrophobic tail with the water. In order to accomplish this the hydrophobic tails align themselves along side each other that, in some cases lead to the formation of a spherical structure called a micelle.

The size of these micelles is dependent on the length of the surfactant molecule and range from 10 - 100 billionths of a meter in diameter. Once there are millions of micelles spinning around in the aqueous medium a reactive monomer is introduced which is also hydrophobic.

Monomer molecules want to minimise their interaction with the water and for this reason they travel to the interior of the micelles where the are stable. The micelles will then swell to accommodate the monomer. The final part of the preparation of the nanoparticles involves the addition of initiator molecules that trigger a chain reaction within the micelles core leading to a polymerisation.

Once the core of the micelles have been converted into a solid polymer sphere of approximately 200 - 1000 billionths of a meter in diameter the spheres may be removed from the emulsion and are ready for use.

Once produced these tiny particles appear as a fine powder that can have drug molecules either attached to their surface or absorbed into their core. At this stage the nanoparticles are solid spheres composed of polymer chains of a fixed length.

This chain length is described by the polymer's molecular weight. For a specific polymer, an example being poly(n-butyl cyanoacrylate), the lower the molecular weight the faster it will degrade within the body and thus the faster the entrained drug is released. The specific site within the body to which the nanoparticles will travel is controlled by the surface chemistry of the nanoparticle. Molecules can be tethered to the particles' surface, which will cause selective uptake, by various organs within the body.

ANALYTICAL TECHNIQUES

Advances in analytical techniques have improved our ability to produce and image nanostructures. Scanning Electron Microscopy (SEM) uses an electron beam to image nanoparticles. They cannot be seen using standard optical microscopy because their diameter is smaller than the wavelength of visible light.

SEM however has its disadvantages in that the powerful electron beam used can often vaporise the particles. A new technique is now becoming available that generates an image by passing a stylus back and forth over the surface of the particles. This stylus is much like the one used on a record player but is a few thousand times smaller.

The technique is called Atomic Force Microscopy (AFM) and its magnifying capability is such that it has been compared with the ability to count the grains of sand on a beach on earth, from the moon! At the moment there is a great deal of research being carried out in the area of nanotechnology.

Materials that change their physical characteristics following exposure to stimuli such as light, heat and changes in pH are being developed. It has been proposed that such materials will lead to the first nano-scale robot arms and similar devices. By combining technologies it may be possible to create 'smart' nanoparticles that will be the next step towards true cellular level medicine.

MEDICAL DIAGNOSIS

Techniques in medical diagnosis and treatment have often been the subject of science fiction literature and cinema. The approaches are surprisingly consistent. In Star Trek, for example, the diagnosis of illness and its treatment is generally non-invasive and mostly painless. In 'The Six Million Dollar Man', an individual who has undergone serious accidental damage is given replacement organs and tissues that function as well, or even better than the originals. What was once the stuff of science fiction is now closer to becoming a reality.

Nature operates at the nanoscale, and today we are acquiring an increasingly profound understanding of natural processes at this scale, enabled by a new generation of scientific instruments.

From this knowledge, we are able to design devices that can either directly interact with, or influence, the behaviour of living cells. Nanotechnology has a trump card to play when applied to medicine. At the nanometre scale, materials often exhibit surprisingly different physical, chemical, and biological properties when compared with the same material in bulk form.

The properties of nanoparticles, such as increased chemical activity and the ability to cross tissue barriers, are leading to new drug targeting and delivery techniques. In the future, it is within the realms of possibility that a nanoparticle may be designed to search for, find and destroy a single diseased cell, taking us even closer to realizing the ultimate goal of disease prevention.

Nanotechnology is also making possible other techniques, such as the stimulation of the body's own mechanisms to successfully repair diseased or damaged tissues, replacing the need for transplants and artificial organs.

In the forseeable future, allied to advances in information and communication technologies, nanotechnology as applied to medicine, will lead to advances in remote monitoring and care, where a patient may be treated at home - a less expensive option, and one that is more conducive to a successful medical outcome than treatment in a surgery or hospital.

Continued research into disease processes at the molecular level is essential for the development of nanomedicine, and involves teams of scientists from across 'conventional' disciplines, such as physics, chemistry, surgery and mathematics, as well as those from the 'new' fields of genomics, proteomics, metabolomics, pharmacokinetic modelling and even microscope design.

Challenges exist in training and managing these multidisciplinary teams and partnerships, and importantly, in finding new solutions to the intellectual property issues that presently hinder the speedy commercialization of new knowledge. Research challenges also relate to understanding and modelling the toxicity of engineered nanoparticles, and requires that toxicologists work alongside medical scientists wherever nanoparticles are involved in drug targeting and contrast agent development.

There are technological challenges, too, in the areas of molecular manufacturing, quality assurance and the eventual programmability of nanodevices.

Nanomedicine will also bring its own set of new legal and ethical challenges to be resolved. Cost benefit analyses are also critical. Where money can be best invested for the greatest social and economic benefit needs to be carefully decided. Should investment be made in prevention or treatment; in diseases of the poor or the rich; in long term hospitalization or costly drug development and deployment? Any strategy for nanomedicine must also be influenced by the facts that resources are finite and demands are great.

Within these constraints the health needs of an ageing population need to be managed, cures found for major lifestyle 'killer' diseases, such as cancer, and for the diseases of the less-developed world, such as HIV/ aids, malaria and tuberculosis. In conclusion, nanoscience and nanotechnology are leading to extraordinary new breakthroughs in medicine that were once the stuff of dreams. EuroNanoForum aims to provide a small but representative snapshot of these exciting opportunities.

TRANSPLANTED ORGANS

Science is moving from transplanted organs to implanting of substitute or artificial organs to stimulating the body to do its own repairs. Tissue regeneration is about enabling the body to ultimately regenerate its own diseased or failed organs. The market worldwide for tissue engineered products already worth $18bn, and many hundreds of thousands of people are awaiting transplants worldwide. This has led to an illegal market in donor organs from the poor to the rich, so the development of effective regeneration techniques offers a multitude of benefits.

Natural Way of Healing

Instead of being surgically repaired, transplanted, or even fixed using prosthetics, tissue or organ failure could be solved by implanting natural tissue and organ mimics which can be fully functional from the start, or grow into the required functionality. Nanotechnology helps to recruit the body's natural healthy cells to promote regeneration of tissue on or around a damaged area.

It can also act as a 'scaffold' to provide a framework for developing tissues to latch onto and penetrate. This application of nanotechnology will result in reconstructed tissue and wound treatments that are superior, longer lasting and more acceptable to everyone involved, most notably the patient.

The beauty of the process is that it is arguably the most natural way of healing, as it is the body's own healthy cells which are regenerating - as they were meant to do - albeit with a small push to get started.

Tissue regeneration would be almost unthinkable without development in nanoscience and nanotechnology. It underpins the design of scaffolds on which the cells are grown, which are composed of special nanocomposite materials hat contains cell growth stimulants, and have nanoscale, cell-friendly surface topographies.

There is even the potential for whole organs to be grown to replace those that have failed through disease or old age. Tissue regeneration offers a revolution in healthcare, with huge benefits for doctors who will be using the technology, to the government who will save money, and most importantly to the patients who are the ultimate beneficiaries.

FIND, FIGHT AND FOLLOW TECHNIQUE

A long standing issue for pharmaceutical companies is to

deliver the right amount of drug to the site of disease. The inability to achieve this has means that drugs need to be administered in excessively high doses, increasing the likelihood that patients may suffer from toxic side effects.

Nanotechnology enables drugs to be targeted more effectively to the site of the disease, and activated 'on arrival'. Drug can be prepared in such a way that they can cross epithelial barriers, and differentiate between healthy cells and diseased ones. Very small drug particles (nanoparticles) are coated with special molecules and travel round the body until they reach the disease site.

Once there, the molecules 'decorating' the drug particles recognise the disease site, bind to it, and the drug is then activated.

This targeting means the drug will work efficiently and more effectively, and side-effects dramatically decrease. For example, chemotherapy in cancer usually means healthy cells are also killed off during the attempt to kill the cancerous cells. Targeted drug delivery circumvents this, and offers the potential for a quicker recovery.

Nanotechnology has other benefits for therapy. In nanoparticulate form, drugs are less prone to degradation, meaning they will have a longer shelf-life. Nanoparticluate drugs can be coated to delay their activation, resulting in the drug remaining in the body longer, reducing the frequency of dosage required. Some drugs may be linked to particles which fluoresce when they reach diseased cells. This is a special technique known as find, fight and follow, and is leading to early detection and treatment.

The area of disease can be identified by imaging the fluoresecent particles, which then deliver their payload of drugs, and the effectiveness of the drugs can be then monitored.

In essence, the application of nanotechnology to drug targeting and delivery offers more effective treatments, lower toxicity and side-effects and reduced cost from reduced dosage. Drug delivery and targeting are areas which are already offering real patient benefits.

FUNCTIONS OF CELL

Humans are essentially a collection of cells! The better we understand how cells work, the more likely it is we can create effective cures, understand the cause of disease better and produce better drugs. Nanotechnology is enabling this understanding.

This session is concerned with understanding of cell structure and function and how a cell manifests disease, and the new tools that are being developed to identify early signs of disease at the cellular level, monitor of healthy and diseased cells and the progress of therapy, and measure the effects of different drugs.

Once we know the signals that a cell gives when it is diseased we can look for that signal. Handin-hand with the growing understanding of cell function, is the technology for measuring and identifying changes that indicate disease. The development of 'lab-on-a-chip' is one such technology which has many applications in disease identification.

These include enabling cells to be studied independently of the body, the speedy screening patients for infection, and the analysis of the hereditary propensity to disease. A lab-on-a-chip is a miniature laboratory that uses techniques developed from the electronics industry, that offers speed, sensitivity, accuracy and portability,

Compared with more traditional techniques, which are slow, expensive resource intensive and may be inaccurate as samples deteriorate with time. In the future, this technology is leading to patient-centred medicine, with patients conducting tests in at their own homes, and downloading the results via their computer to a medical monitoring centre.

Lab-on-a-chip technology also has important applications in being able to quickly screen a large population for disease or potentially killer infections, such as avian flu. Lab-on-a-chip techniques and a knowledge of the human genome is leading to pre-symptomatic disease treatment through identify the individual genetic propensities say to breast or colon cancer or heart disease, so appropriate preventative action can be taken.

This technology is also important for the developing world specifically in some African countries where disease is at epidemic rates. In areas where very little money, or even healthcare is available, the provision of cheap, fast-acting LOCs could offer early identification of disease, and enable more appropriate effective and cheaper treatment options. The cost effectiveness of what is basically a tiny microchip is another selling point; it can be mass produced incredibly cheaply.

NANO-BASED THERAPIES

Nanomedicine offers a huge spectrum of benefits for the better treatment of disease – from early diagnosis to better therapies to eventually preventative medicine; from stimulating the body to regenerate its own diseased tissues to the targeting and activating drugs at the site of diseae. It also offers the possibility of developing better drugs faster, that have fewer side effects and are customised to the patient's own genome.

Three particular areas where nanotechnology will have a major impact in relation to cancer; in relation to reducing animal testing, and in the commercial impact of new nano-based therapies:

Cancer treatment is the past has suffered from a lack of understanding the mechanisms of the disease at the cellular level, late diagnosis, and treatments based on attacking the whole body rather than the disease site. Nanotechnology is offering a totally new approach; from a better understanding of the root causes of disease, to quicker diagnosis and science-base treatments. Today, to quote Dr Mauro Ferrari 'we are seeing the tip of the iceberg' in terms of the potential nanotechnology offers for cancer treatment.

Another area where we are using a hammer to crack a nut is in the area of animal experiments for the validation of drugs. This has obviously produced some benefits in the past, but these are relatively small in relation to the large body of research (and cost) as many apparently good results could not be adequately replicated in humans, or produced unexpected side effects.

With a better knowledge of individual responses to chemicals, nanotechnology is leading to a refinement of techniques for drug discovery that are accurate and meaningful for the individual patient, based on the use of living cells.

Nanotechnology is leading to better treatments based on our knowledge of proteomics and genomics, this is leading to a winwin situation for drug companies and patients as well as huge economic benefits for those countries where the State provides healthcare. Drugs are becoming more efficient as they are increasingly designed to treat specific diseases, patients are cured more rapidly, less care is needed to counteract reactions.

FASTER DIAGNOSIS

Nanotechnology is leading to faster diagnosis and more effective treatment, the early signs of disease, and its exact location. One such application uses quantum dots to locate the disease site.

Quantum dots are semiconductor nanoparticles which can glow very brightly under certain criteria. The quantum dots can be attached to chemicals which link to diseased cells. When the link is made, the quantum dots glow. This allows the sites of disease to be accurately pinpointed, and also malignant and non-malignant tumours can be differentiated. Quantum dots may help in identifying cells which may develop problems in the future.

By recognising certain indicators that cannot be detected using conventional technologies, quantum dots can call attention to potential disease sites, leading to the ideal of disease prevention. Improved drug targeting and delivery is another highly promising area of nanomedicine, which is actually happening now.

Coated drugs in nanoparticulate form can be delivered directly to the diseased cells. This means smaller or less frequent doses are required, and toxicity is reduced as the targeted drug works more effectively. Therefore nanomedicine is leading to the reduction of undesirable and costly side-effects of drugs, as the customisation and targeting of treatment would allow for more control over where the drug goes and what it affects in the body.

Nanotechnology is set to revolutionise healthcare through moving us towards preventative, or pre-emptive medicine, with many health and cost benefits, both to the State and the patient. Impacts are especially envisioned for the ageing and other hitherto disadvantaged populations.

LIMITS OF MEDICAL RESEARCH

Limits and barriers on medical research are being rapidly broken down by the convergence of many disciplines, such as nanotechnology, molecular biology with physics, information technology and even, at last, the humanities! Together, this is resulting in rapid advances in new techniques, processes and tools which are completely changing the way we have approached healthcare in the past.

It is not only researchers from different disciplines that are working together. Networks like Nano2Life are bringing private firms, scientists and research institutes together to solve the problems of treating disease. Strong partnerships are being formed that are providing a range of skills across the board, generating disruptive technologies and completely innovative solutions to old problems.

These could include non-invasive diagnostics, individual-specific drugs, identifying disease at the cellular level, faster validation of new drugs, and stimulating the patient's own body to regenerate diseased or failed organs.

This session will examine convergence and its benefits on a European level. It is case studybased, describing networks and their component organizations, and the benefits that have accrued to industry from EU-supported partnerships and initiatives that connect complementary skills from a wide range of disciplines.

NANOCOMPOSITE MATERIALS

Nanotechnology is expected to play an increasing role in the battle against congenital and degenerative diseases. As we live longer, many more diseases become manifest over a larger proportion of the population, such as Alzheimer's and Parkinson's. We are also experiencing an epidemic of lifestyle diseases, from diabetes to cancer to heart and vascular disease.

Firstly, early diagnosis is critical to many of these diseases, which is enabled by the miniaturization, user-friendliness, cheapness and wide availability of diagnostic techniques to such an extent that patients can monitor their own symptoms. Example include self-testing for diabetes and cholesterol levels. The spectrum of possibilities will expand dramatically; test results can be transmitted to a central health bureau, and therapies prescribed remotely.

Working at the nanoscale has produced new hope for disease of the brain, for example. If the therapy is drug-based, drugs in nanoparticulate form can now cross the blood-brain barrier. If the therapy requires regeneration of diseased tissue, new research is enabling the growth of the patient's own cells on nano-inspired scaffolds that can be then implanted into the brain.

Other areas where Nanotechnology offers hope is in the delivery of drugs in a way that emulated the body's own systems. For example, in diabetes, insulin is required by the body at irregular intervals. Using techniques derived from the electronics industry, this need can be monitored using a small, silicon chip-based implant, and insulin delivered only as required.

Electronics combined with nanotechnology is also the basis of better retinal and cochlear implants. Today, new retinal implants have been developed that enable people previously classified as blind to be reclassified as sighted. Present cochlear implants are crude, difficult to insert, awkward to wear and give poor performance, Future implants will use nanocomposite materials, be small and easy to insert and be cheap to produce. This also has very promising implications for the poor.

Medical Imaging

Medical imaging has advanced from a marginal role in

healthcare to becoming an essential tool of diagnostics over the last 25 years. Molecular imaging and image-guided therapy is now a basic tool for monitoring disease and in developing almost all the applications of in-vivo nanomedicine. Originally, imaging techniques could only detect changes in the appearance of tissues when the symptoms were relatively advanced.

Later, contrast agents were introduced to more easily identify and map the locus of disease.

Today, through the application of nanotechnology, both imaging tools and marker/ contrast agents are being dramatically refined towards the end goals of detecting disease as early as possible, eventually at the level of a single cell, and monitoring the effectiveness of therapy. The convergence of nanotechnology and medical imaging opens the doors to a revolution in molecular imaging, leading eventually to the detection of a single molecule or a single cell in a complex biological environment.

Benefits for Healthcare

Nanomedicine offer many benefits for healthcare in the future. It is a complex subject, and additionally suffers from many misconceptions about the potential of Nanotechnology for medicine, arising out of a confusion between science fiction and fact, even amongst highly educated people. Considering governments and institutions the world over are spending many billions on nanotechnology research and development, a small amount of that funding would be beneficially channelled to wider public awareness and engagement.

As with any new technology, there are risks as well as benefits. Too frequently the downsides of an innovation are publicised, out of proportion to the benefits. A survey carried out last year by researchers at the University of North Carolina established that the more people knew about nanotechnology, the more they thought the benefits would greatly outweigh the risks. The opposite was also true.

This is a key point which indicates a wellinformed public is likely to embrace nanotechnology and, whilst remaining wary of some of the risks, if they see it as bringing a major benefit to their lives. To inform the public, so they can make reasoned and reasonable decisions, and what needs to be put in place to provide reassurance that potential risks are receiving the right level of attention and response.

INTERNATIONAL MEDICAL ETHICS

According to a study by the Canadian Programme on Genomics and Global Health (CPGGH) at the University of Toronto Joint Centre for Bioethics (JCB) - a leading international medical ethics think-tank - nanotechnology may provide the means to help developing countries across as spectrum of areas including improvement in water quality, reduction in environmental and early diagnosis and treatment and even prevention of killer diseases such as malaria, tuberculosis and HIV/AIDS.

Benefits of Nanomedicine

Diagnosis of disease: About a quarter of all Africans are infected with AIDS. It kills millions each year and 95per cent of all new cases of AIDS occur in developing countries. Nano-based devices are leading to easy and fast analysis of a range of diseases that can be undertaken cheaply and easily.

Treatment of Disease: Nanomedicine also has the potential to ensure longer shelf life for drugs, easier administration, and lower dosage requirements. New high throughput screening techniques are leading to cheaper drug discovery; and cell based assays should lead to faster introduction of these drugs into the population.

Sources of Disease: Nanotechnology-enabled systems have application for water purification through membrane filters, which would have the knock-on effect of reducing infection.

Other: Apart from benefiting medicine directly, nanotechnology is also offering access to cheap renewable energy using new polymer based solar power collectors that are lightweig'nt, cheap and efficient.

Needs of the Ageing Population

Nanotechnology could finally mean that growing old does

not necessarily mean loss of faculties and a rapid reduction in quality of life. Nanotechnology is supporting the rehabilitation of the infirm and elderly through the development of intelligent learning prosthetic devices. These include retinal and cochlear implants. Through the development of nanocomposites, nanoelectronics, nanosensors, nanotransducers and biomaterials, the next generation of implants will be more effective, body friendly and less traumatic to insert.

They will reinvigorate the faculties of patients whose hearing or sight has been affected by accident, disease or age. Organ replacement is increasingly common across the generations. Transplantation is fraught with dangers such as rejection and side effects from drugs. As discussed under Tissue Engineering, research is leading to the growth of new organs from a patient's own tissues. The ultimate objective is to produce innovative scaffold materials which canm be seede with the patient's own cells, which, when implanted will regenerate bone, cartilage and skin tissues in the most natural way possible.

Medical textiles is an exciting field of innovation. 'Smart clothes'are being developed which will regulate temperature and even monitor the health of the wearer. This remarkable new technology poses even more questions regarding ethics – the ethics of who should benefit, if there is a choice to be made, where the line should be drawn, if any, at enhancing human performance, and whether new technology will lead to a further widening of the gap between rich and poor, advantaged and disadvantaged.

IMPLANTS AND DEVICES

Nanotechnology, new materials and nanoelectronics are the basis of novel retinal and cochlear implants. New retinal implants have been developed that enable people previously classified as blind (through disease, such as macular degeneration) to be reclassified as sighted.

Present cochlear implants are crude, difficult to insert, awkward to wear and give poor performance, Future implants will use nanocomposite materials, be small and easy to insert and be cheap to produce. This also has very promising implications for the poor.

Over $30bn is spent on implants yearly (hip, knee, dental etc.). Nanotechnology can help bring about better implants. For example, Proteins at the body-implant interface play a decisive role in the acceptance or rejection of the implant. Using nanotechnology we can produce 'bodyfriendly' coatings and also create topographies that cells like, to increase the likelihood of acceptance.

These biocompatible coatings are even being used in devices like hearing aids,reducing the likelihood of irritation or skin reaction. Around 1,000,000 hip and knee-joint replacements are carried out in the EU each year. However, the lifespan of the implants is only around 10 years, shorter if the patient is particularly active or overweight. This poses both a quality of life problem, and a cost effectiveness issue. Revisionary surgery adds about €520m a year to the EU's medical costs.

NANOSENSORS AND DIAGNOSTICS

In nanodiagnostics, the ultimate goal is to identify disease at the earliest stage possible, ideally at the level of a single cell. When testing for disease, nanotechnology offers better sensitivity, specificity and reliability. Several steps from sample preparation to detection can be integrated into a single, tiny device.

An important application is in cheap, reliable devices, based on nanotechnology, that can be used at the "point of care" in clinics by non-expert technicians, and eventually by individuals themselves. Advances in in-vivo diagnostics will rely mainly on imaging. In molecular imaging, the goal is to create sensitive, reliable agents that can detect, deliver and monitor therapy. This is the "find, fight and follow" concept, sometimes also known as *theranostics*.

The likely diseased tissue is firstly imaged, using target-specific contrast agents. These contrast agents can be combined with a drug; and the results of therapy can then be monitored, again using imaging techniques. The advances in diagnostics have presented major opportunities for healthcare, including a plethora of new personalised treatments and early diagnosis for individuals suffering from a wide variety of illnesses such as cancer, diabetes, and heart disease.

Nanotechnology has even made a tiny 'nano needle' possible which can provide important information on disease at the level of a single cell. Other types of nanosensors include cantilevers which are sensitive, simple and cheap and able to accurately detect tiny biochemical changes.

Chapter 10

Applications of Nanotechnology

With nanotechnology, a large set of materials and improved products rely on a change in the physical properties when the feature sizes are shrunk. Nanoparticles for example take advantage of their dramatically increased surface area to volume ratio. Their optical properties, e. g. fluorescence, become a function of the particle diameter. When brought into a bulk material, nanoparticles can strongly influence the mechanical properties of the material, like stiffness or elasticity.

For example, traditional polymers can be reinforced by nanoparticles resulting in novel materials which can be used as lightweight replacements for metals. Therefore, an increasing societal benefit of such nanoparticles can be expected. Such nanotechnologically enhanced materials will enable a weight reduction accompanied by an increase in stability and an improved functionality. There are many applications of nano technology, few of them are shown here.

MEDICINE

The biological and medical research communities have exploited the unique properties of nanomaterials for various applications (e. g., contrast agents for cell imaging and therapeutics for treating cancer). Terms such as *biomedical nanotechnology, bionanotechnology,* and *nanomedicine* are used to describe this hybrid field. Functionalities can be added to nanomaterials by interfacing them with biological molecules or structures.

The size of nanomaterials is similar to that of most biological molecules and structures; therefore, nanomaterials can be useful for both in vivo and in vitro biomedical research and applications. Thus far, the integration of nanomaterials with biology has led to the development of diagnostic devices, contrast agents, analytical tools, physical therapy applications, and drug delivery vehicles.

Diagnostics

Nanotechnology-on-a-chip is one more dimension of lab-on-a-chip technology. Biological tests measuring the presence or activity of selected substances become quicker, more sensitive and more flexible when certain nanoscale particles are put to work as tags or labels. Magnetic nanoparticles, bound to a suitable antibody, are used to label specific molecules, structures or microorganisms.

Gold nanoparticles tagged with short segments of DNA can be used for detection of genetic sequence in a sample. Multicolor optical coding for biological assays has been achieved by embedding different-sized quantum dots into polymeric microbeads. Nanopore technology for analysis of nucleic acids converts strings of nucleotides directly into electronic signatures.

Drug Delivery

The overall drug consumption and side-effects can be lowered significantly by depositing the active agent in the morbid region only and in no higher dose than needed. This highly selective approach reduces costs and human suffering. An example can be found in dendrimers and nanoporous materials. They could hold small drug molecules transporting them to the desired location. Another vision is based on small electromechanical systems; NEMS are being investigated for the active release of drugs.

Some potentially important applications include cancer treatment with iron nanoparticles or gold shells. A targeted or personalized medicine reduces the drug consumption and treatment expenses resulting in an overall societal benefit by reducing the costs to the public health system. Nanotechnology is also opening up new opportunities in implantable delivery systems, which are often preferable to the use of injectable drugs, because the latter frequently display first-order kinetics (the blood concentration goes up rapidly, but drops exponentially over time). This rapid rise may cause difficulties with toxicity, and drug efficacy can diminish as the drug concentration falls below the targeted range.

Tissue Engineering

Nanotechnology can help to reproduce or to repair damaged tissue. This so called "tissue engineering" makes use of artificially stimulated cell proliferation by using suitable nanomaterial-based scaffolds and growth factors. Tissue engineering might replace today's conventional treatments like organ transplants or artificial implants. On the other hand, tissue engineering is closely related to the ethical debate on human stem cells and its ethical implications.

CHEMISTRY AND ENVIRONMENT

Chemical catalysis and filtration techniques are two prominent examples where nanotechnology already plays a role. The synthesis provides novel materials with tailored features and chemical properties: for example, nanoparticles with a distinct chemical surrounding (ligands), or specific optical properties. In this sense, chemistry is indeed a basic nanoscience.

In a short-term perspective, chemistry will provide novel "nanomaterials" and in the long run, superior processes such as "self-assembly" will enable energy and time preserving strategies. In a sense, all chemical synthesis can be understood in terms of nanotechnology, because of its ability to manufacture certain molecules. Thus, chemistry forms a base for nanotechnology providing tailor-made molecules, polymers, etcetera, as well as clusters and nanoparticles.

Catalysis

Chemical catalysis benefits especially from nanoparticles, due to the extremely large surface to volume ratio. The application potential of nanoparticles in catalysis ranges from fuel cell to catalytic converters and photocatalytic devices. Catalysis is also important for the production of chemicals.

Platinum nanoparticles are now being considered in the next generation of automotive catalytic converters because the very high surface area of nanoparticles could reduce the amount of platinum required.

However, some concerns have been raised due to experiments demonstrating that they will spontaneously combust if methane is mixed with the ambient air. Ongoing research at the Centre National de la Recherche Scientifique (CNRS) in France may resolve their true usefulness for catalytic applications. Nanofiltration may come to be an important application, although future research must be careful to investigate possible toxicity.

Filtration

A strong influence of nanochemistry on waste-water treatment, air purification and energy storage devices is to be expected. Mechanical or chemical methods can be used for effective filtration techniques.

One class of filtration techniques is based on the use of membranes with suitable hole sizes, whereby the liquid is pressed through the membrane. Nanoporous membranes are suitable for a mechanical filtration with extremely small pores smaller than 10 nm ("nanofiltration") and may be composed of nanotubes. Nanofiltration is mainly used for the removal of ions or the separation of different fluids.

On a larger scale, the membrane filtration technique is named ultrafiltration, which works down to between 10 and 100 nm. One important field of application for ultrafiltration is medical purposes as can be found in renal dialysis. Magnetic nanoparticles offer an effective and reliable method to remove heavy metal contaminants from waste water by making use of magnetic separation techniques.

Using nanoscale particles increases the efficiency to absorb the contaminants and is comparatively inexpensive compared to traditional precipitation and filtration methods.

Some water-treatment devices incorporating

nanotechnology are already on the market, with more in development. Low-cost nanostructured separation membranes methods have been shown to be effective in producing potable water in a recent study.

ENERGY

The most advanced nanotechnology projects related to energy are: storage, conversion, manufacturing improvements by reducing materials and process rates, energy saving (by better thermal insulation for example), and enhanced renewable energy sources.

Reduction of Energy Consumption

A reduction of energy consumption can be reached by better insulation systems, by the use of more efficient lighting or combustion systems, and by use of lighter and stronger materials in the transportation sector. Currently used light bulbs only convert approximately 5per cent of the electrical energy into light. Nanotechnological approaches like light-emitting diodes (LEDs) or quantum caged atoms (QCAs) could lead to a strong reduction of energy consumption for illumination.

Increasing the Efficiency of Energy Production

Today's best solar cells have layers of several different semiconductors stacked together to absorb light at different energies but they still only manage to use 40 per cent of the Sun's energy. Commercially available solar cells have much lower efficiencies (15-20per cent). Nanotechnology could help increase the efficiency of light conversion by using nanostructures with a continuum of bandgaps. The degree of efficiency of the internal combustion engine is about 30-40per cent at the moment.

Nanotechnology could improve combustion by designing specific catalysts with maximized surface area. In 2005, scientists at the University of Toronto developed a spray-on nanoparticle substance that, when applied to a surface, instantly transforms it into a solar collector.

The use of more Environmentally Friendly Energy Systems

An example for an environmentally friendly form of energy is the use of fuel cells powered by hydrogen, which is ideally produced by renewable energies. Probably the most prominent nanostructured material in fuel cells is the catalyst consisting of carbon supported noble metal particles with diameters of 1-5 nm. Suitable materials for hydrogen storage contain a large number of small nanosized pores.

Therefore many nanostructured materials like nanotubes, zeolites or alanates are under investigation. Nanotechnology can contribute to the further reduction of combustion engine pollutants by nanoporous filters, which can clean the exhaust mechanically, by catalytic converters based on nanoscale noble metal particles or by catalytic coatings on cylinder walls and catalytic nanoparticles as additive for fuels.

Recycling of Batteries

Because of the relatively low energy density of batteries the operating time is limited and a replacement or recharging is needed. The huge number of spent batteries and accumulators represent a disposal problem. The use of batteries with higher energy content or the use of rechargeable batteries or supercapacitors with higher rate of recharging using nanomaterials could be helpful for the battery disposal problem.

INFORMATION AND COMMUNICATION

Current high-technology production processes are based on traditional top down strategies, where nanotechnology has already been introduced silently. The critical length scale of integrated circuits is already at the nanoscale (50 nm and below) regarding the gate length of transistors in CPUs or DRAM devices.

Memory Storage

Electronic memory designs in the past have largely relied on the formation of transistors however research into crossbar switch based electronics have offered an alternative using reconfigurable interconnections between vertical and horizontal wiring arrays to create ultra high density memories. Two leaders in this area are Nantero which has developed a carbon nanotube based crossbar memory called Nano-RAM and Hewlett-Packard which has proposed the use of memristor material as a future replacement of Flash memory.

Novel Semiconductor Devices

An example of such novel devices is based on spintronics. The dependence of the resistance of a material (due to the spin of the electrons) on an external field is called magnetoresistance. This effect can be significantly amplified (GMR - Giant Magneto-Resistance) for nanosized objects, for example when two ferromagnetic layers are separated by a nonmagnetic layer, which is several nanometers thick (e. g. Co-Cu-Co).

The GMR effect has led to a strong increase in the data storage density of hard disks and made the gigabyte range possible. The so called tunneling magnetoresistance (TMR) is very similar to GMR and based on the spin dependent tunneling of electrons through adjacent ferromagnetic layers. Both GMR and TMR effects can be used to create a non-volatile main memory for computers, such as the so called magnetic random access memory or MRAM.

In 1999, the ultimate CMOS transistor developed at the Laboratory for Electronics and Information Technology in Grenoble, France, tested the limits of the principles of the MOSFET transistor with a diameter of 18 nm (approximately 70 atoms placed side by side). This was almost one tenth the size of the smallest industrial transistor in 2003 (130 nm in 2003, 90 nm in 2004, 65 nm in 2005 and 45 nm in 2007).

It enabled the theoretical integration of seven billion junctions on a €1 coin. However, the CMOS transistor, which was created in 1999, was not a simple research experiment to study how CMOS technology functions, but rather a demonstration of how this technology functions now that we ourselves are getting ever closer to working on a molecular scale. Today it would be impossible to master the coordinated assembly of a large number of these transistors on a circuit and it would also be impossible to create this on an industrial level.

Novel Optoelectronic Devices

In the modern communication technology traditional analog electrical devices are increasingly replaced by optical or optoelectronic devices due to their enormous bandwidth and capacity, respectively.

Two promising examples are photonic crystals and quantum dots. Photonic crystals are materials with a periodic variation in the refractive index with a lattice constant that is half the wavelength of the light used.

They offer a selectable band gap for the propagation of a certain wavelength, thus they resemble a semiconductor, but for light or photons instead of electrons. Quantum dots are nanoscaled objects, which can be used, among many other things, for the construction of lasers. The advantage of a quantum dot laser over the traditional semiconductor laser is that their emitted wavelength depends on the diameter of the dot. Quantum dot lasers are cheaper and offer a higher beam quality than conventional laser diodes.

Displays

The production of displays with low energy consumption could be accomplished using carbon nanotubes (CNT). Carbon nanotubes be electrically conductive and due to their small diameter of several nanometers, they can be used as field emitters with extremely high efficiency for field emission displays (FED). The principle of operation resembles that of the cathode ray tube, but on a much smaller length scale.

Quantum Computers

Entirely new approaches for computing exploit the laws of quantum mechanics for novel quantum computers, which enable the use of fast quantum algorithms. The Quantum computer will have quantum bit memory space termed qubit for several computations at the same time.

HEAVY INDUSTRY

An inevitable use of nanotechnology will be in heavy industry.

Aerospace

Lighter and stronger materials will be of immense use to aircraft manufacturers, leading to increased performance. Spacecraft will also benefit, where weight is a major factor. Nanotechnology thereby helps to reduce the size of equipments used.

Hang gliders halve their weight while increasing their strength and toughness through the use of nanotech materials. Nanotech is lowering the mass of supercapacitors that will increasingly be used to give power to assistive electrical motors for launching hang gliders off flatland to thermal-chasing altitudes.

Refineries

Using nanotech applications, refineries producing materials such as steel and aluminium will be able to remove any impurities in the materials they create.

Vehicle Manufacturers

Much like aerospace, lighter and stronger materials will be useful for creating vehicles that are both faster and safer. Combustion engines will also benefit from parts that are more hard-wearing and more heat-resistant.

CONSUMER GOODS

Nanotechnology is already impacting the field of consumer goods, providing products with novel functions ranging from easy-to-clean to scratch-resistant. Modern textiles are wrinkle-resistant and stain-repellent; in the mid-term clothes will become "smart", through embedded "wearable electronics". Already in use are different nanoparticle improved products. Especially in the field of cosmetics, such novel products have a promising potential.

Foods

Nanotechnology can be applied in the production,

processing, safety and packaging of food. A nanocomposite coating process could improve food packaging by placing anti-microbial agents directly on the surface of the coated film. Nanocomposites could increase or decrease gas permeability of different fillers as is needed for different products.

They can also improve the mechanical and heat-resistance properties and lower the oxygen transmission rate. Research is being performed to apply nanotechnology to the detection of chemical and biological substances for sensing biochemical changes in foods.

Household

The most prominent application of nanotechnology in the household is self-cleaning or "easy-to-clean" surfaces on ceramics or glasses. Nanoceramic particles have improved the smoothness and heat resistance of common household equipment such as the flat iron.

Optics

The first sunglasses using protective and antireflective ultrathin polymer coatings are on the market. For optics, nanotechnology also offers scratch resistant surface coatings based on nanocomposites. Nano-optics could allow for an increase in precision of pupil repair and other types of laser eye surgery.

Textiles

The use of engineered nanofibers already makes clothes water- and stain-repellent or wrinkle-free. Textiles with a nanotechnological finish can be washed less frequently and at lower temperatures. Nanotechnology has been used to integrate tiny carbon particles membrane and guarantee full-surface protection from electrostatic charges for the wearer. Many other applications have been developed by research institutions such as the Textiles Nanotechnology Laboratory at Cornell University

Cosmetics

One field of application is in sunscreens. The traditional

chemical UV protection approach suffers from its poor long-term stability. A sunscreen based on mineral nanoparticles such as titanium dioxide offer several advantages. Titanium oxide nanoparticles have a comparable UV protection property as the bulk material, but lose the cosmetically undesirable whitening as the particle size is decreased.

ENERGY APPLICATIONS OF NANOTECHNOLOGY

Over the past few decades, the fields of science and engineering have been seeking to develop new and improved types of energy technologies that have the capability of improving life all over the world. In order to make the next leap forward from the current generation of technology, scientists and engineers have been developing Energy Applications of Nanotechnology. Nanotechnology, a new field in science, is any technology that contains components smaller than 100 nanometers. For scale, a single virus particle is about 100 nanometers in width.

An important subfield of nanotechnology related to energy is nanofabrication. Nanofabrication is the process of designing and creating devices on the nanoscale. Creating devices smaller than 100 nanometers opens many doors for the development of new ways to capture, store, and transfer energy.

The inherent level of control that nanofabrication could give scientists and engineers would be critical in providing the capability of solving many of the problems that the world is facing today related to the current generation of energy technologies.

People in the fields of science and engineering have already begun developing ways of utilizing nanotechnology for the development of consumer products. Benefits already observed from the design of these products are an increased efficiency of lighting and heating, increased electrical storage capacity, and a decrease in the amount of pollution from the use of energy. Benefits such as these make the investment of capital in the research and development of nanotechnology a top priority.

CONSUMER PRODUCTS

Recently, previously established and entirely new companies such as BetaBatt, Inc. and Oxane Materials are focusing on nanomaterials as a way to develop and improve upon older methods for the capture, transfer, and storage of energy for the development of consumer products.

Conserve, a product developed by the Dais Analytic Corporation, uses nanoscale polymer membranes to increase the efficiency of heating and cooling systems and has already proven to be a lucrative design.

The polymer membrane was specifically configured for this application by selectively engineering the size of the pores in the membrane to prevent air from passing, while allowing moisture to pass through the membrane. Polymer membranes can be designed to selectively allow particles of one size and shape to pass through while preventing others of different dimensions. This makes for a powerful tool that can be used in consumer products from biological weapons protection to industrial chemical separations.

A New York based company called Applied NanoWorks, Inc. has been developing a consumer product that utilizes LED technology to generate light. Light-emitting diodes or LEDs, use only about 10per cent of the energy that a typical incandescent or fluorescent light bulb use and typically lasts much longer, which makes them a viable alternative to traditional light bulbs. While LEDs have been around for decades, this company and others like it have been developing a special variant of LED called the white LED.

White LEDs consist of semi-conducting organic layers that are only about 100 nanometers in distance from each other and are placed between two electrodes, which create an anode, and a cathode.

When voltage is applied to the system, light is generated when electricity passes through the two organic layers. This is called electroluminescence. The semiconductor properties of the organic layers are what allow for the minimal amount of energy necessary to generate light. In traditional light bulbs, a metal filament is used to generate light when electricity is run through the filament. Using metal generates a great deal of heat and therefore lowers efficiency.

Research for longer lasting batteries has been an ongoing process for years. Researchers have now begun to utilize nanotechnology for battery technology. mPhase Technologies in conglomeration with Rutgers University and Bell Laboratories have utilized nanomaterials to alter the wetting behaviour of the surface where the liquid in the battery lies to spread the liquid droplets over a greater area on the surface and therefore have greater control over the movement of the droplets.

This gives more control to the designer of the battery. This control prevents reactions in the battery by separating the electrolytic liquid from the anode and the cathode when the battery is not in use and joining them when the battery is in need of use.

ECONOMIC BENEFITS

The relatively recent shift toward using nanotechnology with respect to the capture, transfer, and storage of energy has and will continue to have many positive economic impacts on society. The control of materials that nanotechnology offers to scientists and engineers of consumer products is one of the most important aspects of nanotechnology. This allows for an improved efficiency of products across the board.

A major issue with current energy generation is the loss of efficiency from the generation of heat as a by-product of the process. A common example of this is the heat generated by the internal combustion engine. The internal combustion engine loses about 64per cent of the energy from gasoline as heat and an improvement of this alone could have a significant economic impact.

However, improving the internal combustion engine in this respect has proven to be extremely difficult without sacrificing performance. Improving the efficiency of fuel cells through the use of nanotechnology appears to be more plausible by using molecularly tailored catalysts, polymer membranes, and improved fuel storage.

In order for a fuel cell to operate, particularly of the hydrogen variant, a noble-metal catalyst (usually platinum, which is very expensive) is needed to separate the electrons from the protons of the hydrogen atoms. However, catalysts of this type are extremely sensitive to carbon monoxide reactions. In order to combat this, alcohols or hydrocarbons compounds are used to lower the carbon monoxide concentration in the system.

This adds an additional cost to the device. Using nanotechnology, catalysts can be designed through nanofabrication that are much more resistant to carbon monoxide reactions, which improves the efficiency of the process and may be designed with cheaper materials to additionally lower costs.

Fuel cells that are currently designed for transportation need rapid start-up periods for the practicality of consumer use. This process puts a lot of strain on the traditional polymer electrolyte membranes, which decreases the life of the membrane requiring frequent replacement. Using nanotechnology, engineers have the ability to create a much more durable polymer membrane, which addresses this problem. Nanoscale polymer membranes are also much more efficient in ionic conductivity. This improves the efficiency of the system and decreases the time between replacements, which lowers costs.

Another problem with contemporary fuel cells is the storage of the fuel. In the case of hydrogen fuel cells, storing the hydrogen in gaseous rather than liquid form improves the efficiency by 5per cent. However, the materials that we currently have available to us significantly limit fuel storage due to low stress tolerance and costs. Scientists have come up with an answer to this by using a nanoporous styrene material (which is a relatively inexpensive material) that when super-cooled to around -196°C, naturally holds on to hydrogen atoms and when heated again releases the hydrogen for use.

CAPACITORS: THEN AND NOW

For decades, scientists and engineers have been

attempting to make computers smaller and more efficient. A crucial component of computers are capacitors. A capacitor is a device that is made of a pair of electrodes separated by an insulator that each stores an opposite charge.

A capacitor stores a charge when it is removed from the circuit that it is connected to; the charge is released when it is replaced back into the circuit. Capacitors have an advantage over batteries in that they release their charge much more quickly than a battery.

Traditional or foil capacitors are composed of thin metal conducting plates separated by an electrical insulator, which are then stacked or rolled and placed in a casing. The problem with a traditional capacitor such as this is that they limit how small an engineer can design a computer. Scientists and engineers have since turned to nanotechnology for a solution to the problem.

Using nanotechnology, researchers developed what they call "ultracapacitors. "An ultracapacitor is a general term that describes a capacitor that contains nanocomponents. Ultracapacitors are being researched heavily because of their high density interior, compact size, reliability, and high capacitance. This decrease in size makes it increasingly possible to develop much smaller circuits and computers.

Ultracapacitors also have the capability to supplement batteries in hybrid vehicles by providing a large amount of energy during peak acceleration and allowing the battery to supply energy over longer periods of time, such as during a constant driving speed.

This could decrease the size and weight of the large batteries needed in hybrid vehicles as well as take additional stress off the battery. However, as of now, the combination of ultracapacitors and a battery is not cost effective due to the need of additional DC/DC electronics to coordinate the two.

Nanoporous carbon aerogel is one type of material that is being utilized for the design of ultracapacitors. These aerogels have a very large interior surface area and can have its properties altered by changing the pore diameter and distribution along with adding nanosized alkali metals to alter its conductivity. Carbon nanotubes are another possible material for use in an ultracapacitor. Carbon nanotubes are created by vaporizing carbon and allowing it to condense on a surface. When the carbon condenses, it forms a nanosized tube composed of carbon atoms. This tube has a high surface area, which increases the amount of charge that can be stored. The low reliability and high cost of using carbon nanotubes for ultracapacitors is currently an issue of research.

In a study concerning ultracapacitors or supercapacitors, researchers at the Sungkyunkwan University in the Republic of Korea explored the possibility of increasing the capacitance of electrodes through the addition of fluorine atoms to the walls of carbon nanotubes.

Carbon nanotubes are an increasing form of capacitors due to their superb chemical stability, high conductivity, light mass, and their large surface area. These researchers fluorinated single-walled carbon nanotubes (SWCNTs) at high temperatures to bind fluorine atoms to the walls. The attached fluorine atoms changed the non-polar nanotubes to become polar molecules.

This can be attributed to the charge transfer from the fluorine. This created dipole-dipole layers along the carbon nanotube walls. Testing of these fluorinated SWCNTs against normal state SWCNTs showed a difference in capacitance. It was determined that the fluorinated SWCNTs are advantageous in fabricating electrodes for capacitors and improve the wettability with aqueous electrolytes, which promotes the overall performance of supercapacitors.

While this study brought to knowledge a more efficient example of capacitors, little is known about this new supercapacitor, large scale synthesis is lacking and is necessary for any massive production, and preparation conditions are quite tedious in achieving the final product.

THEORY OF CAPACITANCE

Understanding the concept of capacitance can be helpful in understanding why nanotechnology is such a powerful tool for the design of higher energy storing capacitors. A capacitor's capacitance (C) or amount of energy stored is equal to the amount of charge (Q) stored on each plate divided by the voltage (V) between the plates. Another representation of capacitance is that capacitance (C) is approximately equal to the permittivity (å) of the dielectric times the area (A) of the plates divided by the distance (d) between them. Therefore, capacitance is proportional to the surface area of the conducting plate and inversely proportional to the distance between the plates.

Using carbon nanotubes as an example, a property of carbon nanotubes is that they have a very high surface area to store a charge. Using the above proportionality that capacitance (C) is proportional to the surface area (A) of the conducting plate; it becomes obvious that using nanoscaled materials with high surface area would be great for increasing capacitance.

The other proportionality described above is that capacitance (C) is inversely proportional to the distance (d) between the plates. Using nanoscaled plates such as carbon nanotubes with nanofabrication techniques, gives the capability of decreasing the space between plates which again increases capacitance.

HOW NANOTECHNOLOGY IMPACTS OUR LIVES

NANOCOMPOSITES

Researchers at Pacific Northwest National Laboratory have developed a coating process to make sponge-like silica latch onto toxic metals in water. Self-Assembled Monolayers on Mesoporous Supports easily captures such metals as lead and mercury, which are then recovered for reuse or contained in-place forever. PNNL One example of a SAMMS nanocomposite (Self-Assembled Monolayers on Mesoporous Supports).

An hexagonally close-packed cluster of tubular pores (end view) is shown in the foreground. A single pore, in this case coated with a mercaptopropylsiloxy monolayer. A model of one surfactant molcule is also shown.

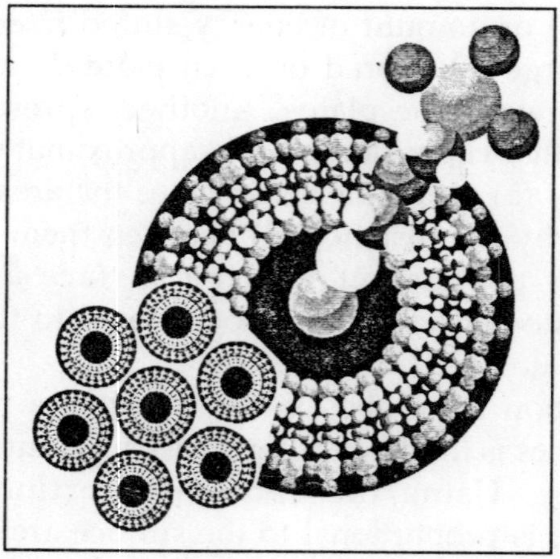

Fig. Self-Assembled Monolayers on Mesoporous Supports

A plastic nanocomposite is being used for "step assists" in the GM Safari and Astro Vans. It is scratch-resistant, light-weight, and rust-proof, and generates improvements in strength and reductions in weight, which lead to fuel savings and increased longevity.

And in 2001, Toyota started using nanocomposites in a bumper that makes it 60 per cent lighter and twice as resistant to denting and scratching.

Impact: Will likely be used on other GM and Toyota models soon, and in other areas of their vehicles, as well as the other auto manufactures, lowering weight, increasing milage, and creating longer-lasting autos. Likely to impact repair shops (fewer repairs needed) and auto insurance companies (fewer claims).

Will also likely soon be seen everywhere weight, weather-proofing, durability, and strength are important factors. Expect NASA, the ESA, and other space-faring organizations to take a serious look, soon, which will eventually result in lower lift costs, which will result in more material being lifted into space.

Strong and light sounds like the perfect recipe for a golf club, of course, which is why so many golf club manufacturers are now devoting big bucks to nano R&D. Thus far, the only company among the big boys to convert research into tangible products is Wilson, which offers three drivers, a fairway wood, four balls and even a golf bag made using nano-materials.

Fig. Electron Micrograph of Typical Silicon Nanocomposite Cross Section Showing Uniform Distribution of Conductive Carbon nanotube network

The Chicago-based firm began looking into the potential of nanotechnology six years ago, but didn't come out with a product - the NCODE series of tennis rackets - until 2004. Later in the year, it started shipping drivers and fairway woods, the Pd5, Dd5, Td5 (MSRP - $300) and FwC (MSRP - $200), whose crowns are constructed with nano carbon which, Wilson claims, creates a low-density, high-strength clubhead. Angus Moir, global business director, says use of nano materials sets Wilson apart from the competition. "They make our products more user friendly," he says.

NANOCRYSTALS

Examples: "Metal nanocrystals might be incorporated into car bumpers, making the parts stronger, or into aluminum, making it more wear resistant. Metal nanocrystals might be used to produce bearings that last longer than their conventional counterparts, new types of sensors and components for computers and electronic hardware.

Nanocrystals of various metals have been shown to be 100 per cent, 200 per cent and even as much as 300 per cent harder than the same materials in bulk form.

Because wear resistance often is dictated by the hardness of a metal, parts made from nanocrystals might last significantly longer than conventional parts. "

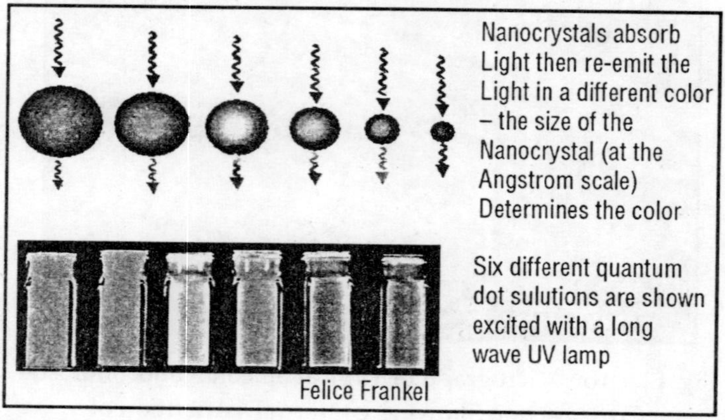

Fig. Nanocrystals Absorb then re-emit the Light in a Different Colour — the Size of the Nanocrystal (in the Angstrom scale) Determines the Colour

Six different quantum dot solutions are shown, excited with a long-wave UV lamp.

Quantum dots are molecular-scale optical beacons. Qdot nanocrystals behave like molecular LEDs (light emitting diodes) by "lighting up" biological binding events with a broad palette of applied colors.

Smith & Nephew markets an antimicrobial dressing covered with nanocrystalline silver (A patented Technology of NUCRYST Pharmaceuticals). The nanocrystalline coating of silver rapidly kills a broad spectrum of bacteria in as little as 30 minutes.

Nanocrystals are an ideal light harvester in photovoltaic devices. They absorb sunlight more strongly than dye molecules or bulk semiconductor material, therefore high optical densities can be achieved while maintaining the requirement of thin films. Perfectly crystalline CdSe nanocrystals are also an artificial reaction centre, separating the electron hole pair on a femtosecond timescale.

Fluorescent nanocrystals have several advantages over

organic dye molecules as fluorescent markers in biology. They are incredibly bright and do not photodegrade.

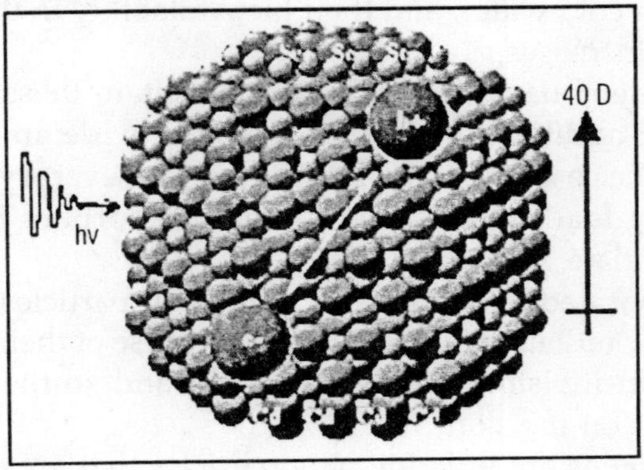

Fig. Nanocrystals

Drug-conjugated nanocrystals attach to the protein in an extracellular fashion, enabling movies of protein trafficking. They also form the basis of a high-throughput fluorescence assay for drug discovery. "

NANOPARTICLES

Stain-repellent Eddie Bauer Nano-Care khakis, with surface fibers of 10 to 100 nanometers, uses a process that coats each fibre of fabric with "nano-whiskers. "Developed by Nano-Tex, a Burlington Industries subsidiary. Dockers also makes khakis, a dress shirt and even a tie treated with what they call "Stain Defender", another example of the same nanoscale cloth treatment.

Dry cleaners, detergent and stain-removal makers, carpet and furniture makers, window covering makers.

BASF's annual sales of aqueous polymer dispersion products amount to around $1. 65 billion. All of them contain polymer particles ranging from ten to several hundred nanometers in size. Polymer dispersions are found in exterior paints, coatings and adhesives, or are used in the finishing of paper, textiles and leather.

Nanotechnology also has applications in the food sector. Many vitamins and their precursors, such as carotinoids, are insoluble in water. However, when skillfully produced and formulated as nanoparticles, these substances can easily be mixed with cold water, and their bioavailability in the human body also increases.

Many lemonades and fruit juices contain these specially formulated additives, which often also provide an attractive colour. In the cosmetics sector, BASF has for several years been among the leading suppliers of UV absorbers based on nanoparticulate zinc oxide.

Incorporated in sun creams, the small particles filter the high-energy radiation out of sunlight. Because of their tiny size, they remain invisible to the naked eye and so the cream is transparent on the skin.

Sunscreens are utilizing nanoparticles that are extremely effective at absorbing light, especially in the ultra-violet (UV) range. Due to the particle size, they spread more easily, cover better, and save money since we use less. And they are transparent, unlike traditional screens which are white. These sunscreens are so successful that by 2001 they had captured 60per cent of the Australian sunscreen market.

Makers of sunscreen have to convert to using nanoparticles. And other product manufactures, like packaging makers, will find ways to incorporate them into packages to reduces UV exposure and subsequent spoilage. The $480B packaging and $300B plastics industries will be directly effected.

Using aluminum nanoparticles, Argonide has created rocket propellants that burn at double the rate. They also produce copper nanoparticles that are incorporated into automotive lubricant to reduce engine wear.

AngstroMedica has produced a nanoparticulate-based synthetic bone. "Human bone is made of a calcium and phosphate composite called Hydroxyapatite. By manipulation calcium and phosphate at the molecular level, we have created a patented material that is identical in structure and composition to natural bone. This novel synthetic bone can be used in areas where natural bone is damaged or removed, such as in the in the treatment of fractures and soft tissue injuries.

NANOSTRUCTURED MATERIALS

Nanodyne makes a tungsten-carbide-cobalt composite powder (grain size less than 15nm) that is used to make a sintered alloy as hard as diamond, which is in turn used to make cutting tools, drill bits, armor plate, and jet engine parts. Every industry that makes parts or components whose properties must include hardness and durability.

Kodak is producing OLED colour screens (made of nanostructured polymer films) for use in car stereos and cell phones. OLEDs (organic light emitting diodes) may enable thinner, lighter, more flexible, less power consuming displays, and other consumer products such as cameras, PDAs, laptops, televisions, and other as yet undreamt of applications.

All current makers of CRTs, liquid crystal displays (LCDs), and other display types.

NANOCLAYS AND NANOCOMPOSITES

Used in packaging, like beer bottles, as a barrier, allowing for thinner material, with a subsequently lighter weight, and greater shelf-life.

$480B packaging and $300B plastics industries. Reduced weight means transportation costs decline. Changing from glass and aluminum - think beer and soda bottles - to plastic reduces production costs. Nanoclays help to hold the pressure and carbonation inside the bottle, increasing shelf life. It is estimated that beer in these containers will gain an extra 60 days (from 120 to 180) of shelf life, reducing spoilage, and decreasing overall costs to the end user.

Nanocor is one company producing nanoclays and nanocomposites, for a variety of uses, including flame retardants, barrier film (as in juice containers), and bottle barrier. "They are not only used to improve existing products, but also are extending their reach into areas formerly dominated by metal, glass and wood. "

NANOCOMPOSITE COATINGS

Wilson Double Core tennis balls have a nanocomposite coating that keeps it bouncing twice as long as an old-style ball. Made by InMat LLC, this nanocomposite is a mix of butyl rubber, intermingled with nanoclay particles, giving the ball substantially longer shelf life. Tires are the next logical extension of this technology: it would make them lighter (better millage) and last longer (better cost performance).

NANOTUBES

Nanoledge makes carbon nanotubes for commercial uses, of which one mundane (marketing tactic) use is in a tennis racket, made by Babolat. The yoke of the racket bends less during ball impact, improving the player's performance.

Once companies like Nanoledge can scale-up their production from grams, to pounds, to tons, and can do so while controlling the type of nanotube they produce, the world becomes their oyster: everywhere strength and weight are a factor - such as in the aerospace, automobile, and airplane industries - they will make a major (disruptive) impact.

Applied Nanotech recently demonstrated a 14" monochrome display based on electron emission from carbon nanotubes. Once the process is perfected, costs will go down, and the high-end market will start being filled. Shortly thereafter, and hand-in-hand with the predictable drop in price of CNTs, production economies-of-scale will enable the costs to drop further still, at which time we will see nanotube-based screens in use everywhere CRTs and view screens are used today.

NANOCATALYSTS

China's largest coal company (Shenhua Group) has licensed technology from Hydrocarbon Technologies that will enable it to liquify coal and turn it into gas. The process uses a gel-based nanoscale catalyst, which improves the efficiency and reduces the cost.

"If the technology lives up to its promise and can economically transform coal into diesel fuel and gasoline, coal-rich countries such as the U. S., China and Germany could depend far less on imported oil. At the same time, acid-rain pollution would be reduced because the liquefaction strips coal of harmful sulfur. "One of the characteristic properties of all nanoparticles has been used from the outset in the manufacture of automotive catalytic converters: The surface area of the particles increases dramatically as the particle size decreases and the weight remains the same.

A variety of chemical reactions take place on the surface of the catalyst, and the larger the surface area, the more active the catalyst. Nanoscale catalysts thus open the way for numerous process innovations to make many chemical processes more efficient and resource-saving – in other words more competitive.

NANOFILTERS

Argonide Nanomaterials, an Orlando based manufacturer of nanoparticles and nanofiltration products, makes a filter that is capable of filtering the smallest of particles. The performance is due to it's nano size alumina fibre, which attracts and retains sub-micron and nanosize particles.

This disposable filter retains 99. 9999+per cent of viruses at water flow rates several hundred times greater than virus-rated ultra porous membranes. It is useful for sterilization of biological, pharmaceutical and medical serums, protein separation, collector/concentrator for biological warfare detectors, and several other applications.

In the future, for one application, sterilizing drinking water, this product may have an impact on so-called Third World peoples, who only have access to dubious sources of water. These are just a few of the many ways in which nanotechnology is working itself into our everyday lives. At present, there are no nanobots, no molecular-scale machines, and no assemblers - these are still in the basic research stages, and may not be seen for decades (although many would argue that a concerted effort would bring them to fruition in just a few years).

AC ELECTROKINETICS: APPLICATIONS FOR NANOTECHNOLOGY

The phenomena of dielectrophoresis and electrorotation,

collectively referred to as AC electrokinetics, have been used for many years to study, manipulate and separate particles on the cellular (1 mm or more) scale.

However, the technique has much to offer the expanding field of nanotechnology, that is the precise manipulation of particles on the nanometre scale. The principles of AC electrokinetics for particle manipulation, review the current state of AC Electrokinetic techniques for the manipulation of particles on the nanometre scale, and consider how these principles may be applied to nanotechnology.

There are two principal approaches to manufacturing on the molecular scale: self-assembly of machines from basic chemical building blocks (the "bottom up" approach) and assembly by manipulating components with much larger devices (the "top down" approach). Whilst the former is considered to be an ideal through which nanotechnology will ultimately be implemented, the latter approach more readily achievable using current technology.

Examples of this include the manipulation of single xenon atoms on a silicon surface and the trapping of single 3nm colloidal particles from solution using electrostatic methods. Another group of methods with potential for use in top-down manufacturing — collectively referred to as AC electrokinetics — will be discussed.

AC Electrokinetic techniques such as dielectrophoresis and electrorotation have been utilised for many years for the manipulation, separation and analysis of cellular-scale particles. The phenomenon occurs due to the interaction of induced dipoles with electric fields, and can be used to exhibit a variety of motions including attraction, repulsion and rotation by changing the nature of the dynamic field.

In many ways, these forces may be viewed as an electrostatic equivalent to optical tweezers and optical spanners in that they exert translational and rotational forces on a body due to the interaction between a body and an imposed field gradient.

Recent advances in semiconductor manufacturing technology have enabled researchers to develop electrodes for manipulating macromolecules as small as 9kDa using both attractive and repulsive AC electrostatic forces, and to concentrate 14nm beads from solution. Trapping of single particles such as viruses and 93nm-diameter latex spheres in contactless potential energy wells has also been demonstrated.

AC Electrokinetics offers advantages over scanning-probe methods of nanoparticle manipulation in that the equipment used is simple, cheap and has no moving parts, relying entirely on the electrostatic interactions between the particle and dynamic electric field.

Furthermore, there is theoretical evidence that as manufacturing technology further improves, single particles considerably smaller than presently studied using AC Electrokinetics may be manipulated.

Ultimately, such a technology will have obvious applications for the manipulation of single molecules. Recent studies have shown that the trapping efficiency of planar electrode arrays is dependent on a number of factors including the magnitude and dimensions of the electric field and the radius of the particle to be trapped. In this paper we will consider the ways in which AC Electrokinetics can benefit nanotechnology, and the constraints on the technique due to factors such as Brownian motion, heating of the medium, and electrode dimensions as the electrode array is miniaturised to the nanometre scale.

DIELECTROPHORESIS

Consider a dielectric particle suspended in a spatially non-uniform electric field such as that shown in figure. The applied field induces a dipole in the particle; the interaction of the induced dipole with the electric field generates a force. Due to the presence of a field gradient, these forces are not equal and there is a net movement. If the particle is more conductive than the medium around it, the dipole aligns with the field and the force acts up the field gradient towards the region of highest electric field.

If the particle is less polarisable than the medium, the dipole aligns against the field and the particle is repelled from regions of high electric field. The force is dependent on the induced dipole, and is unaffected by the direction of the electric field, responding only to the field gradient.

Since the alignment of the field is irrelevant, this force can also be generated in AC fields which has the advantage of reducing any *electrophoretic* force (due to any net particle charge) to zero.

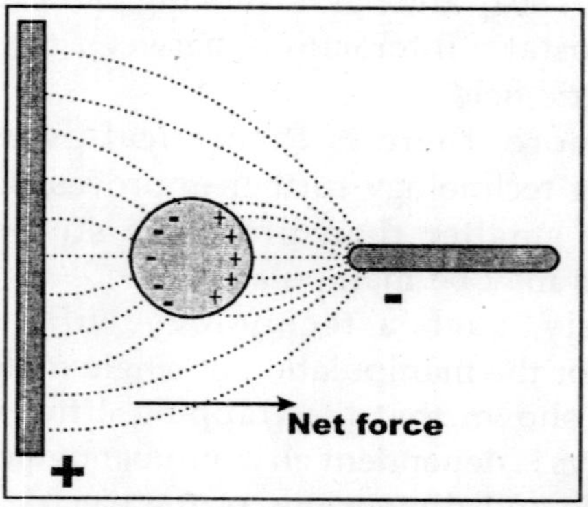

Fig. A schematic of a Polarisable Particle Suspended within a Point-Plane Electrode System

When the particle polarises, the interaction between the dipolar charges with the local electric field produces a force. Due to the inhomogeneous nature of the electric field, the force is greater in the side facing the point than that on the side facing the plane, and there is net motion towards the point electrode. This effect is called positive dielectrophoresis. If the particle is less polarisable than the surrounding medium, the dipole will align counter to the field and the particle will be repelled from the high field regions, called negative dielectrophoresis.

This effect was first termed Dielectrophoresis by Pohl. The dielectrophoretic force, F_{DEP}, acting on a homogeneous, isotropic dielectric sphere, is given by:

$$F_{DEP} = 2\pi r^3 \varepsilon_m \operatorname{Re}[K(\omega)] \nabla E^2$$

where r is the particle radius, ε_m is the permittivity of the suspending medium, ∇ is the Del vector operator, E is the *rms.*

electric field and *Re* [K (ω)] the real part of the Clausius-Mossotti factor, given by:

$$K(\omega) = \frac{\varepsilon_p^* - \varepsilon_m^*}{\varepsilon_p^* + 2\varepsilon_m^*}$$

where ε_m^* and ε_p^* are the complex permittivities of the medium and particle respectively, and $\varepsilon' = \varepsilon - \frac{j\sigma}{\omega}$ with σ the conductivity, ε the permittivity and ω the angular frequency of the applied electric field.

The frequency-dependence of *Re* [K (ω)] indicates that the force on the particle varies with the frequency. The magnitude of *Re* [K (ω)] also varies depending on whether the particle is more or less polarisable than the medium. If *Re* [K (ω)] is positive, then particles move to regions of highest field strength (positive dielectrophoresis); the converse is negative dielectrophoresis where particles are repelled from these regions. By careful construction of the electrode geometry which creates the electric field, it is possible to create electric field morphologies so that potential energy minima are bounded by regions of increasing electric field strengths.

In such electrodes, particles experiencing positive dielectrophoresis are attracted to the regions of highest electric field (typically the electrode edges, particularly where adjacent electrodes are close), whilst particles experiencing negative dielectrophoresis are trapped in isolated field minima.

ELECTROROTATION

If a polarisable particle is suspended in a rotating electric field, the induced dipole will form across the particle and rotate in synchrony with the field. However, if the angular velocity of the field is sufficiently large, the time taken for the dipole to form (the relaxation time of the dipole) becomes significant and the dipole will lag behind the field.

This results in non-zero angle between field and dipole, which induces a torque in the body and causes it to rotate asynchronously with the field; the rotation can be with or against the direction of rotation of the field, depending of whether the lag is less or more than 180°.

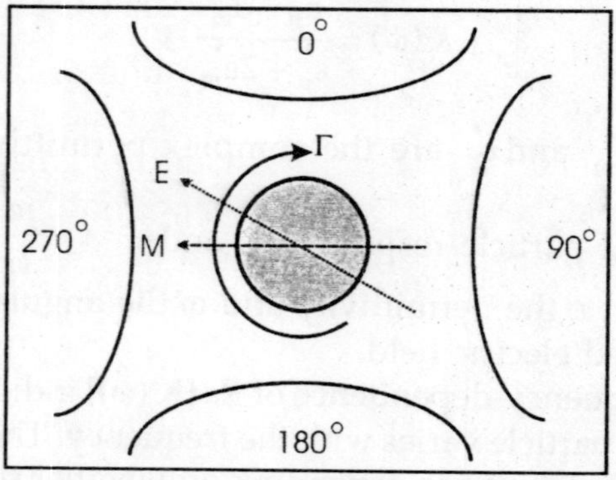

Fig. A schematic of a Polarisable Particle Suspended in a Rotating Electric field Generated by four Electrodes with 90° Advancing Phase

If the electric field E rotates sufficiently quickly, the induced dipole M will lag behind the electric field by an angle related to the time taken for the dipole to form (the relaxation time). The interaction between the electric field and the lagging dipole induces a torque G in the particle, causing the particle to rotate. This effect is known as *electrorotation*.

The general equation for time-averaged torque G experienced by a spherical polarisable particle of radius r suspended in a rotating electric field E is given by:

$$\Gamma = -4\pi\varepsilon_m r^3 \, Im[K(\omega)]E^2$$

where $Im[K(\omega)]$ represents the imaginary component of the Clausius-Mossotti factor. The minus sign indicates that the dipole moment lags the electric field. When viscous drag is accounted for, the rotation rate R(ω) of the particle is given by:

$$R(\omega) = -\frac{\varepsilon_m \, Im[K(\omega)]E^2}{2\eta}$$

Note that there are two significant differences between the expression for torque as shown in equation. Firstly, the relationship with the electric field is as a function of the square of the electric field rather than of the *gradient* of the square of the electric field. Secondly, the torque depends on the imaginary rather than the real part of the Clausius-Mossotti factor.

This is significant in that a particle may experience both dielectrophoresis and electrorotation simultaneously, and the magnitudes and directions of both are related to the interaction between the dielectric properties of particle and medium. As an example of this, a plot of the frequency-dependence of a solid particle exhibiting a single dielectric dispersion is shown in figure.

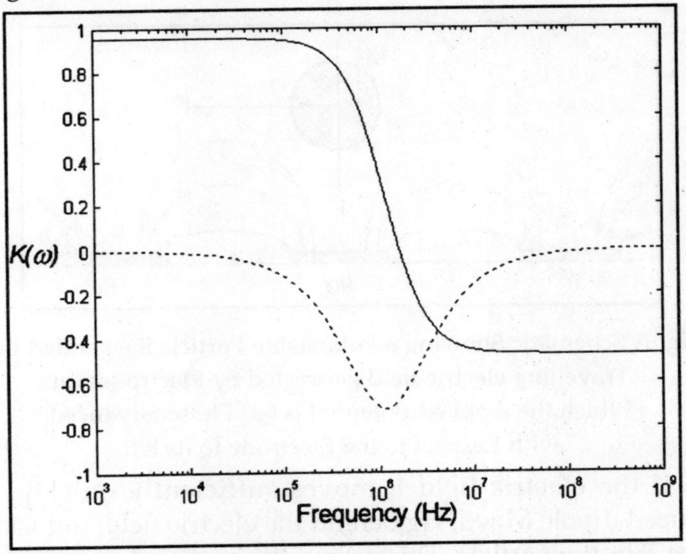

Fig. A plot of the real (solid line) and imaginary (dotted line) parts of the Clausius-Mossotti factor. The magnitude and signs of the real and imaginary parts govern the magnitude and direction of the dielectrophoretic force and electrorotational torque respectively

TRAVELLING-WAVE DIELECTROPHORESIS

Travelling-wave dielectrophoresis is a linear analogue of electrorotation; instead of electrodes being arranged in a circle, they are laid out as "tracks", but the relationship of phases (each electrode being 90° phase advanced from the last)

remains. This produces an electric field wave which "travels" along the electrodes. When this wave interacts with a polarisable particle, a dipole is induced which moved with the electric field peak.

If the travelling wave is moving sufficiently quickly, the dipole will lag behind the field in a similar manner to that observed in electrorotation. However, as the wave is travelling in a linear (rather than rotational) manner, the result is the induction of a force rather than a torque, with particles moving along the electrodes like a train travelling along tracks. This is demonstrated schematically in figure.

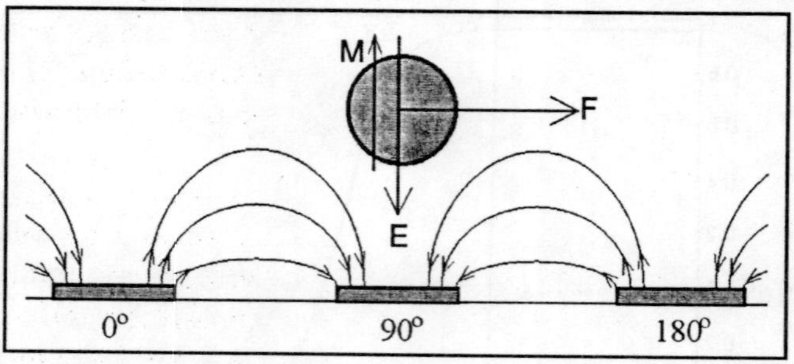

Fig. A Schematic Showing a Polarisable Particle Suspended in a Travelling electric field generated by Electrodes on Which the Applied Potential is 90° Phase-advanced with Respect to the Electrode to its left

If the electric field E moves sufficiently quickly, the induced dipole M will lag behind the electric field, inducing a force F in the particle. This causes the particle to move along the electrodes, a phenomenon known as *travelling-wave dielectrophoresis*.

The value of the force F_{TWD} is given by equation:

$$F_{TWD} = \frac{-4\pi\varepsilon_m \gamma^3 Im[K(\omega)]E^2}{\gamma}$$

where l is the wavelength of the travelling wave. This phenomenon was first observed by Batchelder. However, it remained largely unexplored until work by Masuda *et al.* on using *synchronously* induced motion in low frequency travelling fields that travelling wave dielectrophoresis became significant. This work led to Fuhr *et al.* effectively rediscovering asynchronous travelling wave dielectrophoresis. Since then, a large corpus of study has been conducted on this technique, including theoretical studies and demonstrations of practical devices for electrostatic pumping and biotechnological applications.

ELECTRODE CONFIGURATIONS

In practice, dielectrophoretic manipulation of particles is performed on planar electrode arrays. Such electrode arrays employ patterned electrodes fabricated of gold or a similar conductor deposited on glass or silicon. A common electrode configuration used in AC Electrokinetics research is a quadrupole arrangement where four electrodes point towards a central enclosed region, as shown in the schematic in figure.

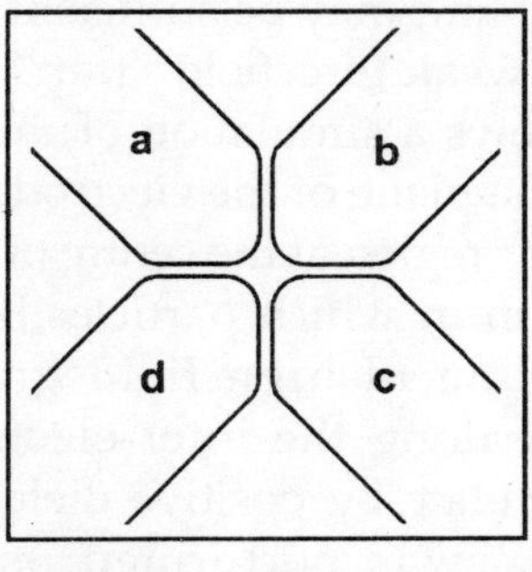

Fig. A Schematic of Typical Quadrupole Electrode Micros -tructures used in Dielectrophoresis Experiments

The electrodes are typically fabricated of gold on a glass microscope slide. The gap between opposing electrodes in the centre of the array is typically of the order 10-50 μm across, but can be as small as 500nm or as large as 1mm. To induce dielectrophoretic motion in particles suspended near the electrode array, electrodes would be energised such that a and c are of the same phase, and b and d are in antiphase to them. To cause electrorotation in particles within the central gap, electrodes b, c and d would be phase-shifted by 90°, 180° and 270° with respect to electrode a.

These electrodes may be of a variety of designs, according to the degree of field non-uniformity required. The advantage with this arrangement is twofold. Firstly, electrodes can easily be used for dielectrophoresis, electrorotation or both. By energising the electrodes such that the phase angle of applied AC field between adjacent electrodes is 180°, particles will experience only dielectrophoresis.

If the phase angle is reduced such that adjacent electrodes differ by 90°, the particle will experience both electrorotation and dielectrophoresis. The second advantage of the quadrupole design is that at the centre of the electrode array, there exists a well-defined electric field minimum, which is surrounded by well-defined field maxima at the electrode edges.

This definition is advantageous when attempting to manipulate nanometre-sized particles, where Brownian motion is significant and may be sufficient to enable particles to "escape" from a weak forcefield "trap". This is illustrated in figure, which shows a simulation of the electric field in a plane 5 μm above the plane of the electrodes shown in figure.

The dark circular region at the centre of the electrode array is the field minimum in which particles become trapped by the surrounding ring of high field gradient. The field maximum extends along the inter-electrode gaps, where particles will accumulate by positive dielectrophoresis.

The simulation was performed using a Method of Moments software suite developed by the author at the University of Wales at Bangor. As can be seen, there is an electric field minimum (dark region) at the centre of the electrode array, surrounded by a ring of high electric field gradient.

Particles experiencing negative dielectrophoresis are repelled into this minimum and become trapped. The electric field strength is high (white regions) along the electrode edges, where particles collect by positive dielectrophoresis.

Advances in fabrication technology such as the use of electron beam lithography have allowed the manufacture of electrodes with feature sizes of the order 100 nm, covering many square cm. By reducing the size of inter-electrode gaps, electric fields of 10^7 Vm^{-1} can be generated by applying 10V across a 1 μm inter-electrode gap.

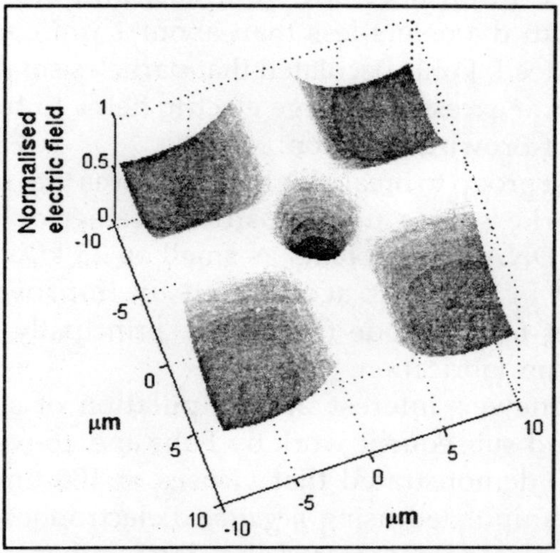

Fig. A Simulation of the Electric Field in the Plane 5 μm above the Electrode array

More advanced dielectrophoretic traps have been developed in which two sets of electrodes were stacked so as to form an octopole field "cage". This type of geometry has allowed the controlled manipulation of populations of particles in all 3 dimensions.

APPLICATIONS TO NANOTECHNOLOGY

Concentration of Colloids from Solution

Since the force experienced by a particle undergoing dielectrophoresis scales as a function of the particle volume, it was believed for many years that a lower threshold of particle size existed, below which the dielectrophoretic force would be overcome by Brownian motion.

It was held that to increase the force would require electric fields of such magnitude that local medium heating would increase local fluid flow, again acting to prevent dielectrophoretic manipulation. As electrode fabrication techniques were relatively crude, generating electric fields of sufficient magnitude required large potentials to be applied across large inter-electrode volumes, and consequently particles with diameters less than about 1 μm could not be trapped. Indeed, Pohl speculated that particles smaller 500nm would require excessively large electric fields to trap against the action of Brownian motion.

The first group to break this threshold was that of Washizu and co-workers who used positive dielectrophoresis to precipitate DNA and proteins as small as 25 kDa. This step downward in size was accelerated by improvements in technologies for electrode fabrication, principally the use of electron beam fabrication.

This renewed interest in manipulation of sub-micron particles, and subsequent work by Fuhr and co-workers and Green *et al.* demonstrated that viruses of 100 nm diameter could be manipulated using *negative* dielectrophoresis.

It was also demonstrated that latex spheres of 14 nm diameter could be trapped by either positive or negative dielectrophoresis, as demonstrated by Müller *et al.* and reproduced here in figure.

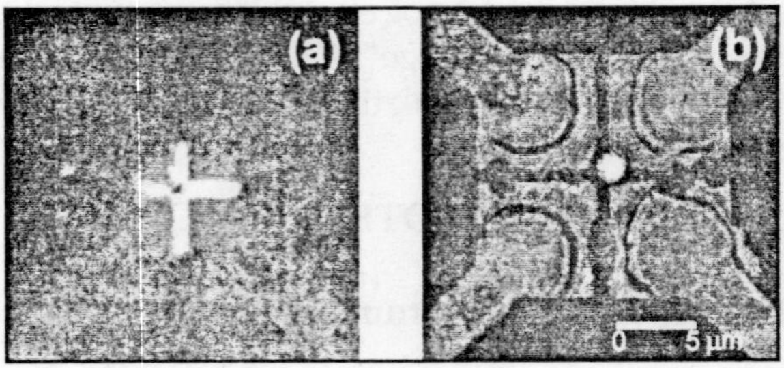

Fig. A Photograph of 14 nm Diameter, Fluorescently Labelled Latex Spheres Precipitated from an Aqueous Solution (Conductivity Adjusted to 2. 5 mSm^{-1} by Addition of KCl) by Dielectrophoresis

Subsequent work by Hughes and co-workers demonstrated that by varying the frequency of the applied electric field, Herpes viruses can be trapped using either positive or negative electric fields. Another study demonstrated that molecules of the 68 kDa-protein avidin can be concentrated from solution by both positive and negative dielectrophoresis.

The electrodes were of similar construction to those shown in figure, with 500 nm between adjacent electrodes and 2 μm across the inter-electrode gap. In photograph (a), the spheres collect in the high-field regions between the electrode edges when a $8V_{pp}$, 2MHz signal is applied.

In photograph (b) the spheres collect in a ball at the centre of the trap, which is levitated a few μm above the electrode plane when the applied signal is changed to $10V_{pp}$, 10MHz. Collection of particles is rapid and begins immediately after the field is applied. This experiment was performed at the Humboldt University of Berlin, reproduced with permission. Taken from Müller *et al*.

The most obvious nanotechnological application of this technique is in the concentrating of "parts" of molecular machinery to one site. This could be either in the form of bringing together of two interlocking molecules in a highly dilute solution, or collecting "fuel" for a nanomachine from solution.

Furthermore, by attracting microspheres and colloidal gold particles which have been functionalised using antibodies, Velev and Kaler have used dielectrophoresis to construct microscopic biosensors by "stacking" particles of different types, which bind on contact.

Manipulation of Single Nanoparticles

In 1996, Müller *et al*. Demonstrated that a single 1 μm latex sphere could be trapped by negative dielectrophoresis, and subsequently repeated the experiment with a latex sphere of 650 nm diameter.

Further work by Hughes and Morgan demonstrated that single Herpes viruses, viral capsids and latex spheres as small as 93 nm diameter could be held in stable traps using negative dielectrophoresis, as shown in figure. This study also considered in some detail the mathematical conditions for a particle to be trapped by either positive or negative dielectrophoresis, concluding that for stable trapping, an approximate minimum particle radius r is given by the expression:

$$r > \sqrt[3]{\frac{10kT}{\pi\varepsilon_m \Delta d Re[K(\omega)]\nabla E^2}}$$

where k is Bolzmann's constant, T is the temperature, and $\Delta\, d$ is the thickness of the electric field barrier of magnitude, ∇E^2. Since the factors $\Delta\, d$ and ∇E^2 are related to the size of the electrode array, they place limits on the minimum size of particle an array can be used to manipulate.

Studies by Hughes *et al.* indicated that for electrode structures with 5Vp-p applied to electrodes with 6 μm between opposing electrodes and 2 μm between adjacent electrodes, the minimum stable trapping radius for single particle is approximately 30 nm.

This research indicated that the key variable governing $\Delta\, d$ is the size of the centre of the trap, which in that research was an area of approximately 5 μm radius centred on the middle of the electrodes.

Therefore it should theoretically be possible to reduce the whole array to this size; this would benefit the use of this technique in nanotechnology since the manipulator would be of the order of a few microns across.

Furthermore, since $\Delta\, d$ scales linearly with distance and ∇E^2 scales as the inverse cube of distance, equation indicates that reducing the size of the trap by a factor x whilst maintaining the same activating potentials will increase the efficiency of the trap by reducing the minimum trappable particle radius by a factor $x^{2/3}$.

An electrode structure similar to that shown in figure, with 6 μm between opposing electrode tips and 2 μm between adjacent electrodes. The medium conductivity was adjusted to $5mSm^{-1}$ by addition of EDTA, and the applied signal was $5V_{pp}$, 15 MHz.

The bead occupies the space at the point of the "spike" at the centre of the electrode structure. The vertical dimension was extracted from light emission data from fluorescence micrographs. This experiment was performed at the University of Glasgow, reproduced with permission. Due to the fact that this trapping is contactless, with particles being repelled from the electrode edges, particles may move upwards and will eventually leave the trap. In order to prevent this happening closed field cages, such as those fabricated by Schnelle *et al.* must be employed.

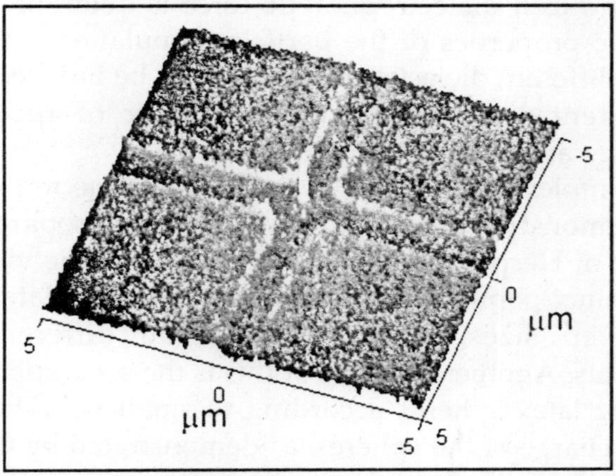

Fig. A pseudo-3D image of a Single 93nm Latex Sphere Suspended in a Trap by Negative Dielectrophoresis

An octopole cage of sufficiently small dimension could in theory allow full 3D manipulation of single particles of less than 30 nm diameter, and has been demonstrated to function for aggregates of 14 nm particles. The technology to produce such electrodes for single-molecule manipulation exists, but the principal problem to be overcome is the observation of the trapped particle.

An alternative approach would be to use an AFM tip to generate field non-uniformity, and use positive dielectrophoresis to attract particles to it. These could be positioned by the AFM and then "released" by changing the frequency of the signal to the AFM tip so as to repel the particles from it.

The principle behind this — using positive dielectrophoresis to attract particles to a 3D, microscopic needle tip —

has already been demonstrated to coat silicon tips with nanodiamond.

PARTICLE SEPARATION

In 1992 Gascoyne *et al.* demonstrated that positive and negative dielectrophoresis could be used to separate a heterogeneous population of normal and leukaemic murine erythrocytes into two separate populations. Since the direction of the induced dielectrophoretic force is determined by the dielectric properties of the particle, populations of particles having different dielectric properties can be induced to move in different directions by careful choice of applied field frequency and medium conductivity.

Examples of this work in the submicrometre range have been demonstrated by Morgan *et al.* in the separation of a mixture of Herpes Simplex and Tobacco Mosaic viruses into two distinct populations and the separation of latex spheres of different sizes, or of similar size but different surface treatments. Another example of this is the separation of 93nm diameter latex spheres according to small variations in the surface charge of the spheres, as demonstrated by Green and Morgan.

In a nanotechnological context, the dielectrophoretic separation of particles allows for the manufacture of a tuneable electrostatic filter. By altering the frequency of the applied electric field, particles for two populations can be either both attracted to the electrode array, both repelled from it, or one can be attracted whilst the other is repelled.

Since the effect is volumetric — repulsion acts for several microns into the solution, according to the magnitude of the electric field — the result is to alter the local concentration of the two particle types dramatically.

PARTICLE TRANSPORT

Whilst dielectrophoresis is useful for attracting particles to specific points, there are applications which require the moving of particles between one position and another. This requirement may be to relocate particles for processing, for separation purposes, or to gather in particles from a larger volume than can be achieved by conventional dielectrophoresis alone. Two means of particle transport have been demonstrated in the literature.

Firstly, conventional dielectrophoresis has been used to move 100 nm latex spheres in a thermal ratchet, where dielectrophoresis is used to rectify the motion of particles induced by Brownian motion. The induced motion can be either towards or against the ratchet potential, according to whether particles experience positive or negative dielectrophoresis, or the time between activation of ratcheting pulses. This technique has the advantage that since only two phases are required, the electrodes for inducing such motion are easy to fabricate.

The alternative, and more popular, means of transporting particles using AC electrokinetics is by the application of travelling-wave dielectrophoresis. Whilst the majority of work on travelling-wave dielectrophoresis has concerned micrometre-sized objects such as blood cells, some work has been performed on the concentration of particles on a surface using so-called "meander" electrodes. These structures use four electrodes in a series of interlocking spirals to generate a travelling wave; at the centre of the spiral, the electrodes form a quadrupole-type electrode array.

It has been demonstrated that by careful manipulation of the amplitudes of the potentials on these electrode structures, it is possible to "steer" the motion of particles across the array.

Tools such as these could be used as the basis for "conveyor belts" for *factories on a chip*, wherein different chemical processes may be carried out of the same chip, with operations performed by electrostatic or chemical means and the resultant output transferred to a new process by AC Electrokinetic means.

ELECTROROTATION

Electrorotation was first examined scientifically in the early 1960s but was most fully explored by Arnold and Zimmermann, who established the technique for direct measurement of the value of *K(w)* by measuring the time taken for a particle to rotate.

This application has little bearing to nanotechnology, not least due to the difficulty of seeing such small particles rotate, but electrorotation may be useful as a means of inducing rotation in molecular machinery.

Recent work by Berry and Berg has involved using electrorotation to drive the molecular motor of *E. Coli* bacteria backwards at speeds of up to 2,000 Hz.

Since the electrodes involved in that work were applying 10 V across inter-electrode gaps of 50 μm or more, considerably higher fields (and hence induced torques) could be applied with electrodes with inter-electrode gaps of the order of 1 μm. If one considers that electrorotation is an electrostatic equivalent of laser spanners, then there exists a number of applications within current nanotechnological thought to which electrorotation might be applied.

There are a number of advantages that favour electrorotation over its optical equivalent; most notable advantages are that electrorotation-induced torque is easily controlled by altering the frequency of the electric field, and that there does not need to be a direct optical path to the part to be manipulated. Electrorotation has been demonstrated for the induction of torque in micron-scale motors by Hagedorn et al.

Brownian Motion, Conduction and Convection

Aside from the engineering of electrodes small enough to generate large electric field non-uniformity, there are other factors that present problems with scaling AC Electrokinetic techniques from the micrometre to nanometre scale. These include Brownian motion induced by thermal noise in the surrounding medium and effects of conduction and convection of the medium induced by the high electric fields used for manipulating colloidal particles.

For sufficiently high concentrations of particles, a diffusion force would also be significant; however, the majority of work on dielectrophoresis has used sufficiently low concentrations of particles for each particle to be regarded as single and isolated.

Brownian motion due to thermal noise is well documented and a number of studies have considered the force required to trap particles against Brownian motion. Experimental studies with proteins have demonstrated that Brownian motion does not prevent the trapping particles as small as 68 kDa (equivalent to a cube 6 nm along a face) by negative dielectrophoresis, or 25 kDa by positive dielectrophoresis.

Recent studies on the forces involved in dielectrophoretic manipulation of sub-micrometre particles examined the balance of Brownian motion and dielectrophoretic forces by considering the amount of time required to determine, with 99. 7per cent confidence (3 standard deviations) that an observed force is due to dielectrophoresis.

Any force applied to a particle will ultimately result in what Ramos *et al.* describes as "observable deterministic motion", but if the force is small the time taken to observe it may be of the order of years!

However, the forces used in the experiments described in the literature are sufficiently large for the applied force to observably overcome Brownian motion to be of the order of a second or (often significantly) less. It may ultimately be possible to manipulate single small particles precisely using dielectrophoretic forces, but this will be more difficult to achieve for very small colloids and work needs to be done to study this.

Other disruptive effects are caused by the presence of very high electric fields covering small areas of the electrode array, creating electrohydrodynamic (EHD) motion in the fluid. An in-depth study of this subject has been conducted by Ramos *et al.*, and presented in summary form by Green *et al.* In these works, the authors demonstrated that there are two major EHD forces which are significant in driving fluid flow around electrodes, those being electrothermal and electroosmotic.

The former force is due to localised heating of the medium causing discontinuities of medium conductivity and permittivity, which was found to be insignificant due to the small volumes these discontinuities occupy (as the electrodes are so small). The latter force is due to the interaction of the tangential electric field with the diffuse double layer above the electrode surfaces.

This creates a fluid pumping action, the magnitude of which can be several orders of magnitude larger than the dielectrophoretic force. However, careful electrode design can allow the EHD forces to be used as an aid to particle manipulation.

The techniques of AC Electrokinetics, principally dielectrophoresis and electrorotation, have much to offer the expanding science of nanotechnology. Whereas dielectrophoretic methods have been applied to the manipulation of objects in the micro-scale, recent work shows that the technique is applicable to the manipulation of *nano*-scale particles. This opens a wide range of potential applications for AC electrokinetics to the development of mainstream nanotechnology.

The fundamental challenge in the advancement of nanotechnology is the development of precision tools for the manipulation of macromolecules in solution phase. The techniques presented in this paper go some way to addressing this by providing tools for the trapping, manipulation and separation of biomolecules and small latex spheres using tools (electrodes) on the micrometre scale.

There exist a number of avenues down which AC electrokinetics can be developed to meet the needs of nanotechnology more readily.

These include the design of electrodes for trapping of smaller single particles, increasing the precision to which particles may be trapped, and examining the extent to which particles may be manoeuvred within the trap by alteration of the electric field.

NANOTECHNOLOGY AND MEDICINE

Disease and ill health are caused largely by damage at the molecular and cellular level. Today's surgical tools are, at this scale, large and crude. From the viewpoint of a cell, even a fine scalpel is a blunt instrument more suited to tear and injure than heal and cure. Modern surgery works only because cells have a remarkable ability to regroup, bury their dead and heal over the injury.

Nanotechnology, "the manufacturing technology of the 21st century," should let us economically build a broad range of complex molecular machines (including, not incidentally, molecular computers). It will let us build fleets of computer controlled molecular tools much smaller than a human cell and built with the accuracy and precision of drug molecules.

Such tools will let medicine, for the first time, intervene in a sophisticated and controlled way at the cellular and molecular level. They could remove obstructions in the circulatory system, kill cancer cells, or take over the function of subcellular organelles. Just as today we have the artifical heart, so in the future we could have the artificial mitochondrion.

Equally dramatic, nanotechnology will give us new instruments to examine tissue in unprecedented detail. Sensors smaller than a cell would give us an inside and exquisitely precise look at ongoing function. Tissue that was either chemically fixed or flash frozen could be analyzed literally down to the molecular level, giving a completely detailed "snapshot" of cellular, subcellular and molecular activities.

There is broad agreement (though not consensus) that we will at some point in the future be able to inexpensively fabricate essentially any structure that is consistent with chemical and physical law and specified in molecular detail.

The most direct route to achieving this capability involves positioning and assembling individual atoms and molecules in a fashion conceptually similar to snapping together LEGO blocks. By designing and building programmable self replicating manufacturing systems that incorporate these principles we should be able to achieve very low manufacturing costs. While the design and development of such programmable self replicating molecular manufacturing systems will be a major task and will likely require many years or a few decades, it appears that this kind of capability, to quote Feynman. Design concepts for general purpose self replicating manufacturing systems have been discussed for many years, and their utility in manufacturing has been emphasized recently.

These proposals draw on a body of work started by von Neumann. A wide range of methods have been considered. The von Neumann architecture for a self replicating system is the ancestral and archetypal proposal.

THE VON NEUMANN ARCHITECTURE

Von Neumann's proposal consisted of two central elements: a universal computer and a universal constructor. The universal computer contains a programme that directs the behaviour of the universal constructor. The universal constructor, in turn, is used to manufacture both another universal computer and another universal constructor. Once construction is finished the programme contained in the original universal computer is copied to the new universal computer and programme execution is started.

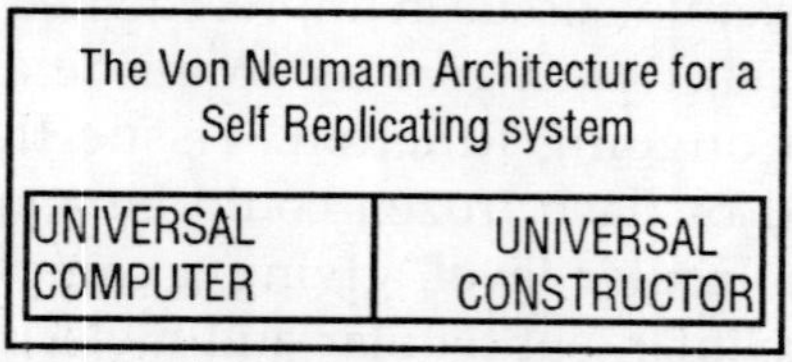

Von Neumann worked out the details for a constructor that worked in a theoretical two-dimensional cellular automata world (parts of his proposal have since been modeled computationally. The constructor had an arm which it could move about and which could be used to change the state of the cell at the tip of the arm. By progressively sweeping the arm back and forth and changing the state of the cell at the tip, it was possible to create "objects" consisting of regions of the two-dimensional cellular automata world which were fully specified by the programme that controlled the constructor.

While this solution demonstrates the theoretical validity of the idea, von Neumann's kinematic constructor (which was not worked out in such detail) has had perhaps a greater influence, for it is a model of general manufacturing which can more easily be adapted to the three-dimensional world in which we live. The kinematic constructor was a robotic arm which moved in three-space and which grasped parts from a sea of parts around it. These parts were then assembled into another kinematic constructor and its associated control computer.

An important point to notice is that self replication, while important, is not by itself an objective. A device able to make copies of itself but unable to make anything else would not be very valuable.

Von Neumann's proposals centered around the combination of a universal constructor, which could make anything it was directed to make, and a universal computer, which could compute anything it was directed to compute.

It is this ability to make any one of a broad range of structures under flexible programmatic control that is of value. The ability of the device to make copies of itself is simply a means to achieve low cost, rather than an end in itself.

DREXLER'S ARCHITECTURE FOR AN ASSEMBLER

Drexler's assembler follows the von Neumann kinematic architecture, but is specialized for dealing with systems made of atoms. The emphasis here (in contrast to von Neumann's proposal) is on small size. The computer and constructor both shrink to the molecular scale, while the constructor takes on additional detail consistent with the desire to manipulate molecular structures with atomic precision. The molecular constructor has two major subsystems:

- A positional capability
- The tip chemistry.

The positional capability might be provided by one or more small robotic arms, or alternatively might be provided by any one of a wide range of devices that provide positional control. The emphasis, though, is on a positional device that is very small in scale: perhaps 0. 1 microns (100 nanometers) or so in size.

The tip chemistry is logically similar to the ability of the von Neumann universal constructor to alter the state of a cell at the tip of the arm, but now the change in "state" corresponds to a change in molecular structure. That is, we must specify a set of well defined chemical reactions that take place at the tip of the arm, and this set must be sufficient to allow the synthesis of the structures of interest.

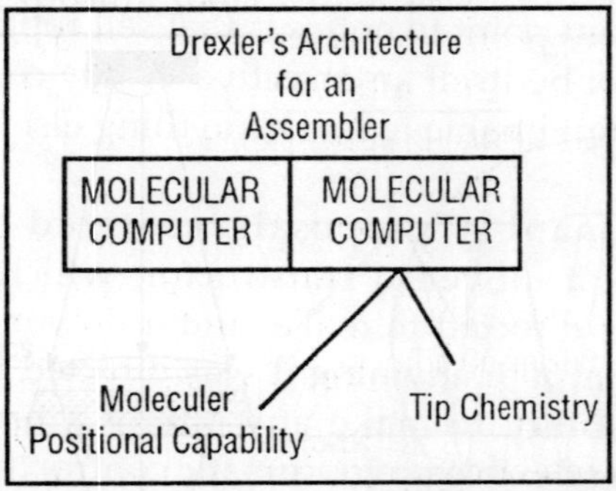

It is worth noting that current methods in computational chemistry are sufficient to model the kinds of structures that will appear in a broad class of molecular machines, including all of the structures and reactions needed for some assemblers.

SIZE OF DEVICES

Drexler's proposal for molecular mechanical logic is the most compact and, from the system point of view, the best worked out. The logic elements ("locks," roughly the equivalent of a single transistor) need occupy a volume of only a few cubic nanometers. Even including system overhead (power, connections, etc). the volume per element should still be less than 100 cubic nanometers. A 10,000 element logic system (enough to hold a small processor) would occupy a cube no more than 100 nanometers on a side.

That is, a volume only slightly larger than 0. 001 cubic microns would be sufficient to hold a small computer. This compares favorably with the volume of a typical cell (thousands of cubic microns) and is even substantially smaller than subcellular organelles. Operating continuously at a gigahertz such a computer would use less than 10^{-9} watts. By comparison, the human body uses about 100 watts at rest and more during exercise. Slower operation and the use of reversible logic would reduce power consumption, quite possibly dramatically.

A variety of molecular sensors and actuators would also fit in such a volume. A molecular "robotic arm" less than 100 nanometers long should be quite feasible, as well as molecular binding sites 10 nanometers in size or less.

By contrast, a single red blood cell is about 8 microns in diameter (over 80 times larger in linear dimensions than our 100 nanometer processor). Devices of the size range suggested above (~0. 1 microns) would easily fit in the circulatory system and would even be able to enter individual cells.

KILLING CANCER CELLS

Given such molecular tools, we could design a small device able to identify and kill cancer cells. The device would have a small computer, several binding sites to determine the concentration of specific molecules, and a supply of some poison which could be selectively released and was able to kill a cell identified as cancerous.

The device would circulate freely throughout the body, and would periodically sample its environment by determining whether the binding sites were or were not occupied. Occupancy statistics would allow determination of concentration. Today's monoclonal antibodies are able to bind to only a single type of protein or other antigen, and have not proven effective against most cancers.

The cancer killing device suggested here could incorporate a dozen different binding sites and so could monitor the concentrations of a dozen different types of molecules. The computer could determine if the profile of concentrations fit a pre-programmed "cancerous" profile and would, when a cancerous profile was encountered, release the poison.

Beyond being able to determine the concentrations of different compounds, the cancer killer could also determine local pressure. A pressure sensor little more than 10 nanometers on a side would be sufficient to detect pressure changes of less than 0. 1 atmospheres (a little over a pound per square inch. One atmosphere is ~10^5 Pascals, so PV in this case would be $(0.1 \times 10^5) \times (10^{-8})^3$ or $10^4 \times 10^{-24}$ or 10^{-20} joules.

Multiple samples would be required to achieve reliable operation, as kT is ~4×10^{-21} joules at body temperature. Linear increases in sensor volume would produce exponential increases in immunity to thermal noise and linear improvements in pressure sensitivity if that were to prove useful. Doubling the linear dimensions of the sensor would produce an eight-fold increase in both volume and pressure sensitivity).

As acoustic signals in the megahertz range are commonly employed in diagnostics (ultrasound imaging of pregnant women, for example), the ability to detect such signals would permit the cancer killer to safely receive broadcast instructions.

By using several macroscopic acoustic signal sources, the cancer killer could determine its location within the body much as a radio receiver on earth can use the transmissions from several satellites to determine its position (as in the widely used GPS system). Megahertz transmission frequencies would also permit multiple samples of the pressure to be taken from the pressure sensor, as the CPU would be operating at gigahertz frequencies.

The cancer killer could thus determine that it was located in (say) the big toe. If the objective was to kill a colon cancer, the cancer killer in the big toe would not release its poison. Very precise control over location of the cancer killer's activities could thus be achieved.

The cancer killer could readily be reprogrammed to attack different targets (and could, in fact, be reprogrammed via acoustic signals transmitted while it was in the body). This general architecture could provide a flexible method of destroying unwanted structures (bacterial infestations, etc).

PROVIDING OXYGEN

A second application would be to provide metabolic support in the event of impaired circulation. Poor blood flow, caused by a variety of conditions, can result in serious tissue damage. A major cause of tissue damage is inadequate oxygen. A simple method of improving the levels of available oxygen despite reduced blood flow would be to provide an "artificial red blood cell. "We will consider a simple design here: a sphere with an internal diameter of 0. 1 microns (100 nanometers) filled with high pressure oxygen at ~1,000 atmospheres (about 10^8 pascals).

The oxygen would be allowed to trickle out from the sphere at a constant rate. Diamond has a Youngs modulus of about 10^{12} pascals. An atomically precise diamondoid structure should be able to tolerate a stress of greater than 5 x 10^{10} pascals (5per cent of the modulus). Thus, a 0. 1 micron sphere of oxygen at a pressure of 10^*8* pascals could be contained by a hollow diamondoid sphere with an internal diameter of 0. 1 microns and a thickness of less than one nanometer.

This thickness, thin as it is, results in an applied stress on the diamond of well under 1per cent of its modulus — from a purely structural point of view we should be able to use a very large "bucky ball," i. e., a sphere whose surface is a single layer of graphite. Perhaps the most complex issue involved in the selection of the material is the reaction of the body's immune system. While some suitable surface structure should exist which does not trigger a response by the immune system — after all, there are many surfaces in the body that are not attacked — the selection of a specific surface structure will require further research.

To give a feeling for the range of possible surface structures, the hydrogenated diamond surface could have a variety of "camouflauge" molecules covalently bound to its surface. A broad range of biological molecules could be anchored to the surface, either directly or via polymer tethers.

The Van der Waals' equation of state is $(p+a/v^2)\ (v-b) = RT$, where p is the pressure, v is the volume per mole, R is the universal gas constant, T is the temperature in Kelvins, and a and b are constants specific to the particular gas involved. For oxygen, a = 1. 36 atm liter^2/mole^2 and b = 0. 03186 liter/mole and R = 0. 0820568 liter-atmospheres/mole-kelvin. A mole of oxygen at 1,000 atmospheres and at body temperature (310

Kelvins) occupies 0. 048 liters, or about 21 moles/liter. A mole of oxygen at 1 atmosphere and 310 Kelvins occupies 25. 4 liters, or about 0. 04 moles/liter.

This implies a compression of ~530 to 1. A resting human uses ~240 cc/minute of oxygen, so a liter of oxygen compressed to 1,000 atmospheres should be sufficient to maintain metabolism for about 36 hours (a day and a half). It might be desirable to replace less than a liter of blood with our microspheres of compressed oxygen, but it should still be quite feasible to provide oxygen to tissue even when circulation is severely compromised for periods of at least many hours from a single infusion.

Controlled release of oxygen from the diamondoid sphere could be done using the selective transport method proposed by Drexler. Transport in the "wrong" direction, but simply reversing the direction of rotor motion would result in transport from inside the reservoir to the external fluid. By driving a rotor at the right speed, oxygen could be released from the internal reservoir into the external environment at the desired rate.

More sophisticated systems would release oxygen only when the measured external partial pressure of oxygen fell below a threshold level, and so could be used as an emergency reserve that would come into play only when normal circulation was interupted.

Full replacement of red blood cells would involve the design of devices able to absorb and compress oxygen when the partial pressure was above a high threshold (as in the lungs) while releasing it when the partial pressure was below a lower threshold (as in tissues using oxygen). In this case, selective transport of oxygen into an internal reservoir would be required.

If a single stage did not provide a sufficiently selective transport system, a multi-staged or cascaded system could be used. Compression of oxygen would presumably require a power system, perhaps taking energy from the combustion of glucose and oxygen (thus permitting free operation in tissue). Release of the compressed oxygen should allow recovery of a significant fraction of the energy used to compress it, so the total power consumed by such a device need not be great.

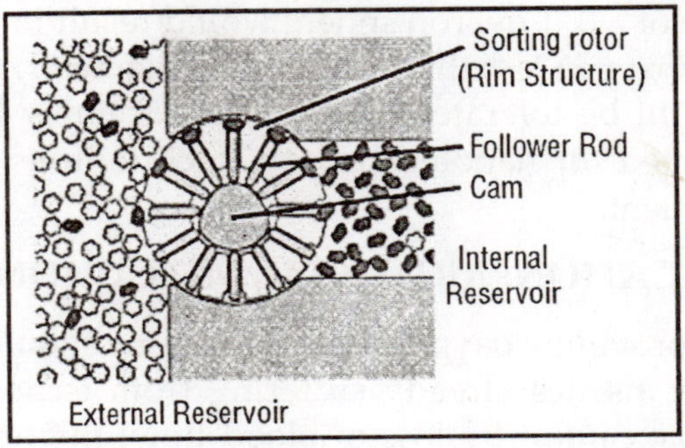

Fig. Rotor motion

If the device were to simultaneously absorb carbon dioxide when it was present at high concentrations (in the tissue) and release it when it was at low concentrations (in the lungs), then it would also provide a method of removing one of the major products of metabolic activity. Calculations similar to those given above imply a human's oxygen intake and carbon dioxide output could both be handled for a period of about a day by about a liter of small spheres.

As oxygen is being absorbed by our artificial red blood cells in the lungs at the same time that carbon dioxide is being released, and oxygen is being released in the tissues when carbon dioxide is being absorbed, the energy needed to compress one gas can be provided by decompressing the other. The power system need only make up for losses caused by inefficiencies in this process.

These losses could presumably be made small, thus allowing our artificial red blood cells to operate with little energy consumption.

By comparison, a liter of blood normally contains ~0. 2 liters of oxygen, while one liter of our spheres contained ~530 liters of oxygen (where "liter of oxygen" means, as is common in the literature on human oxygen consumption, one liter of the gas under standard conditions of temperature and pressure). Thus, our spheres are over 2,000 times more efficient per unit volume than blood; taking into account that blood is only about half occupied by red blood cells, our spheres are over 1,000 times more efficient than red blood cells.

Failure of a 0. 1 micron sphere would result in creation of a bubble of oxygen less than 1 micron in diameter. Occasional failures could be tolerated. Given the extremely low defect rates projected for nanotechnology, such failures should be very infrequent.

AN APPLICATION: ARTIFICIAL MITOCHONDRIA

While providing oxygen to healthy tissue should maintain metabolism, tissues already suffering from ischemic injury (tissue injury caused by loss of blood flow) might no longer be able to properly metabolize oxygen. In particular, the mitochondria will, at some point, fail. Increased oxygen levels in the presence of nonfunctional or partially functional mitochondria will be ineffective in restoring the tissue.

However, more direct metabolic support could be provided. The direct release of ATP, coupled with selective release or absorption of critical metabolites, should be effective in restoring cellular function even when mitochondrial function had been compromised. The devices restoring metabolite levels, injected into the body, should be able to operate autonomously for many hours (depending on power requirements, the storage capacity of the device and the release and uptake rates required to maintain metabolite levels).

FURTHER POSSIBILITIES

While levels of critical metabolites could be restored, other damage caused during the ischemic event would also have to be dealt with. In particular, there might have been significant free radical damage to various molecular structures within the cell, including its DNA. If damage was significant restoring metabolite levels would be insufficient, by itself, to restore the cell to a healthy state.

Various options could be pursued at this point. If the cellular condition was deteriorating (unchecked by the normal homeostatic mechanisms, which presumably would cease to function when cellular energy levels fell below a critical value), some general method of slowing further deterioration would be desirable.

Cooling of the tissue, or the injection of compounds that would slow or block deteriorative reactions would be desirable. As autonomous molecular machines with externally provided power could be used to restore function, maintaining function in the tissue itself would no longer be critical. Deliberately turning off the metabolism of the cell to prevent further damage would become a feasible option. Following some interval of reduced (or even absent) metabolic activity during which damage was repaired, tissue metabolism could be restarted again in a controlled fashion.

It is clear that this approach should be able to reverse substantially greater damage than can be dealt with today. A primary reason for this is that autonomous molecular machines using externally provided power would be able to continue operating even when the tissue itself was no longer functional. We would finally have an ability to heal injured cells, instead of simply helping injured cells to heal themselves.

NANOTECHNOLOGY AND MEDICAL RESEARCH

Advances in medical technology necessarily depend on our understanding of living systems. With the kind of devices discussed earlier, we should be able to explore and analyse living systems in greater detail than ever before considered possible.

Autonomous molecular machines, operating in the human body, could monitor levels of different compounds and store that information in internal memory. They could determine both their location and the time. Thus, information could be gathered about changing conditions inside the body, and that information could be tied to both the location and the time of collection. Physical samples of small volumes (nano tissue samples) could likewise be taken.

These molecular machines could then be filtered out of the blood supply and the stored information (and samples) could be analyzed. This would provide a picture of activities within healthy or injured tissue. This new knowledge would give us new insights and new approaches to curing the sick and healing the injured.

SNAPSHOTS OF THE ENTIRE SYSTEM

More dramatically, it should be feasible to take "snapshots" of tissue samples and analyse the structure down to the molecular level. First, a small tissue sample could be either fixed or frozen. Chemical fixation can be used to rapidly block most tissue changes. Ultra fast freezing of small tissue samples is an effective method of halting essentially all chemical processes and diffusion of all molecules.

Once fixed or frozen, the tissue sample could be analyzed in a leisurely fashion. With nanotechnology (and indeed, to some extent with current STM and AFM technologies, though rather more expensively) it should be feasible to scan the tissue surface in molecular detail, and store that information in a computer.

Once the surface had been scanned, it could be removed in a very selective and precise fashion, and scanned again. As an example, the use of a positionally controlled carbene has been proposed for use in the synthesis of complex diamondoid structures. Such a positionally controlled carbene is highly reactive and, if positioned at an appropriate site on the surface of the tissue being analyzed, would readily react with a surface molecule. This surface molecule could then be removed.

A wide variety of other "sticky" molecular tools could be brought up to the surface and allowed to react with surface molecules, which could then be removed, exposing the layers beneath. The use of a positionally controlled carbene implies that the environment in which it is used must be inert. This requirement could be satisfied by analyzing the tissue sample at very low temperature (a few Kelvins) and in a very good vacuum. Under these conditions the tissue specimen would remain stable during even a protracted analysis process.

While this process can readily be envisioned for very small structures, nanotechnology should make massive parallelism feasible. That is, a single positional device could be used at a certain speed to provide information about a certain (rather small) volume of tissue in a reasonable time.

Nanotechnology should permit the manufacture of a large number of small devices, each able to analyse a small volume. Given enough such devices operating in parallel, larger volumes could be analyzed and the information from many individual devices integrated to provide a coherent picture of the larger whole.

Effective use of this option will require massive computational power — which will also be made feasible with nanotechnology. Estimates of the computational power that should be provided by nanotechnology exceed 10^{24} logic operations per second for a single desktop computer. This amount of raw computational power should make control of a large number of parallel devices feasible, and should permit integration and analysis of the information so obtained.

In short, tissue samples could be "frozen" (either literally by ultrafast cooling or figuratively by chemical fixation) and the entire resulting tissue sample could be analyzed down to the level of individual molecules. The information so obtained could be processed by computers able to handle the flood of data produced.

The resulting "snapshots" will provide us with an instantaneous look at metabolic and cellular activities across even relatively large volumes of tissue. Such an ability should revolutionize our understanding of the complex processes that take place in living systems. The possibility of truly revolutionary advances in our medical abilities has also created renewed interest in cryonics.

The abilities discussed here might well take years or decades to develop. It is quite natural to ask: "When might we see these systems actually used?" The scientifically correct answer is, of course, "We don't know. "That said, it is worth noting that if progress in computer hardware continues as the trend lines of the last 50 years suggest, we should have some form of molecular manufacturing in the 2010 to 2020 time frame. After this, the medical applications will require some additional time to develop.

The remarkably steady trend lines in computer hardware, however, give a false sense that there is a "schedule" and that developments will spontaneously happen at their appointed time. This is incorrect. How long it will take to develop these systems depends very much on what we do. If focused efforts to develop molecular manufacturing and its medical applications are pursued, we will have such systems well within our lifetimes. If we make no special efforts the schedule will slip, possibly by a great deal.

Index

D

E

F

G

H

I